The Politics of Improving Urban Air Quality

To Keith Jones, unsung hero of the midfield.   WG

To the children of my children, Martina, Lukas and Miriam, who hopefully one day will live in more sustainable cities.   PK

To Andrea Banks.   AP

The Politics of Improving Urban Air Quality

Edited by
Wyn Grant
Professor of Politics, University of Warwick, UK

Anthony Perl
Director, Research Unit for Public Policy Studies, University of Calgary, Canada

and

Peter Knoepfel
Professor of Public Policy Analysis and Environmental Policies, Institute for Advanced Studies in Public Administration, University of Lausanne, Switzerland

Edward Elgar
Cheltenham, UK • Northampton, MA, USA

© Wyn Grant, Anthony Perl and Peter Knoepfel 1999

All rights reserved. No part of this publication may be reproduced, stored in a retrieval system or transmitted in any form or by any means, electronic, mechanical, photocopying, recording, or otherwise without the prior permission of the publisher.

Published by
Edward Elgar Publishing Limited
Glensanda House
Montpellier Parade
Cheltenham
Glos GL50 1UA
UK

Edward Elgar Publishing, Inc.
6 Market Street
Northampton
Massachusetts 01060
USA

A catalogue record for this book is available from the British Library

Library of Congress Cataloguing in Publication Data
The politics of improving urban air quality / edited by Wyn Grant, Anthony Perl, Peter Knoepfel.
  Includes bibliographical references (p.   ).
  1. Air quality management—Economic aspects. 2. Air quality management—Government policy. 3. Air—Pollution—Government policy. I. Grant, Wyn. II. Perl, Anthony, 1962–
III. Knoepfel, Peter.
HC79.A4P65 1999
363.739'2—dc21   98–30465
                                                                          CIP

ISBN 1 85898 696 6

Printed and bound in Great Britain by
MPG Books Ltd, Bodmin, Cornwall

Contents

List of Figures vi
List of Tables vii
Preface viii

1 Introduction
 Wyn Grant, Peter Knoepfel and Anthony Perl 1
2 When policy networks collide: the institutional dynamic of air pollution policy-making in two Canadian cities
 Anthony Perl, Jane Hargraft and Kevin Muxlow 13
3 'Soft' institutions for hard problems: instituting air pollution policies in three Italian regions
 Marco Giuliani 31
4 Improving air quality in Italian cities: the outcome of an emergency policy style
 Carlo Desideri and Rudy Lewanski 52
5 Shifting tools and shifting meanings in urban traffic policy: the case of Turin
 Luigio Bobbio and Alberico Zeppetella 73
6 Changing definitions and networks in clean air policy in France
 Corinne Larrue and C.A. Vlassopoulou 93
7 Lyon's urban transportation policy and the air quality problem: a policy network approach
 Grégoire Marlot and Anthony Perl 107
8 Clean air and transport policy in Switzerland: the case of Berne
 Daniel Marek 127
9 Conclusion: institution building for sustainable urban mobility policies
 Peter Knoepfel, Wyn Grant and Anthony Perl 144

Bibliography 168
Index 183

Figures

3.1	Resources and outcomes of air pollution policies	37
3.2	Mobilization of actors with regard to air pollution problems	41
3.3	The geography of cognitive maps	45
5.1	Turin: from Restricted Traffic Zone to parking charges	78
5.2	Turin: traffic survey points	79
5.3	The Prisoner's Dilemma	81
5.4	Individual pay-offs of cooperation and defection	82
5.5	Total pay-offs of cooperation and defection	85
7.1	Lyon's public transport policy network	116
7.2	Lyon's private transport policy network	118
7.3	An integrated transport policy network	125

Tables

3.1	A typology of regional environmental performance	35
3.2	Cognitive maps of policy-makers regarding problem definition	42
3.3	Measures of policy networks	47
4.1	Prevailing mode of passenger transport in Italy, 1970–93	53
4.2	Prevailing mode of freight transport in Italy, 1970–93	53
7.1	Transportation investments in Lyon, 1968–75	112
7.2	Mobility evolution by transportation mode: Lyon, 1976–96	113
7.3	Evolution of commuting trips within the metropolitan area of Lyon, 1976–82	113
7.4	Air pollution in the COURLY, 1990	122
8.1	Overview of all measures examined in the city of Berne	141

Preface

This book arises from the work of a European Commission COST committee, COST-CITAIR, Action 618. COST is the acronym for the French equivalent of 'European Cooperation in the Field of Scientific and Technical Research'; it is principally a framework for research and development cooperation, allowing for the coordination of national research on a European level.

The COST-CITAIR programme as a whole on 'Science and research for better air in European cities' is made up of four COST actions:

C615 Database, monitoring and modelling of urban air pollution
C616 Mobile sources of air pollution in urban areas
C617 Stationary sources of urban air pollution in urban areas
C618 Institution building and information policy in the field of urban air pollution

The focus of the CITAIR programme was not air pollution in general, but the local dimension of the problem and possible local initiatives and abatement strategies. COST 618 was the only action examining problems of urban air pollution from a social science perspective.

The work of the COST 618 Action was guided by Peter Knoepfel from Switzerland as chair, and by Wyn Grant from the UK and Henning Schroll from Denmark as the coordinators of the two working groups on institution building and information policy. The working party on information policy is preparing a separate book.

This volume arises in particular from a workshop held in Rome in October 1996 and hosted by the CNR (Consiglio Nazionale delle Ricerche), Istituto di Studi sulle Regioni. We would particularly like to thank Carlo Desideri for making it possible for us to meet in Rome. This workshop and the work of COST 618 as a whole was funded through DG XII of the European Commission. Funding was provided from the same source for the attendance of Anthony Perl, bringing a valuable Canadian dimension to the work of the action. Finally, the European Commission also funded a scientific mission to Canada for Wyn Grant, which enabled him to work on the editing of the book with Anthony Perl and to meet with air quality experts in Vancouver to discuss the COST-CITAIR programme and share knowledge and experience.

Anthony Perl would like to thank the Social Sciences and Humanities Research Council of Canada (Grant #410-94-0492) for aiding both his research in the chapter with Hargraft and Muxlow (Chapter 2) and his contribution to the overall book. Wyn Grant would like to thank the Cabinet Office for nominating him as a British member of the COST committee. Peter Knoepfel would like to thank the Swiss Agency for the Environment, Forests and Landscapes for its substantial support in starting and implementing the COST 618 action in Switzerland and on the international level.

Students of comparative public policy seek both to advance the theoretical perspectives and methodologies of their area of study and to contribute to helping to solve complex social problems. This book seeks to achieve both those aims.

Wyn Grant
Peter Knoepfel
Anthony Perl

1. Introduction

Wyn Grant, Peter Knoepfel and Anthony Perl

Air pollution from mobile sources (cars and other vehicles) is one of the more pressing and intractable problems facing public policy-makers. The essence of the dilemma is this: the car offers advantages of autonomy, choice, mobility and privacy not readily available from other forms of transport. As countries become more prosperous, one consequence is a rapid growth in car ownership, as South Korea and China illustrate. Similarly, as economies grow road transport is often the cheapest and most convenient means of moving freight. For all the talk of the 'greening' of society and the readiness of respondents in attitude surveys to declare themselves as friends of the environment, the reality in affluent societies is that the convenience and flexibility of the car is the predominant influence on daily routines. Changes in values have not yet had a significant impact on urban travel behaviour.

The growth of road transport brings with it many disbenefits. These include deaths and injuries from motor accidents; problems of noise pollution; and the negative aesthetic effects in both cities and rural areas of major roads. The current underpricing or non-pricing of roads gives a distorted market signal with externality cost estimates ranging from 100 to 300 per cent of user costs (Miller and Moffett, 1993; MacKenzie et al., 1992). One of the major negative consequences of the increased use of road transport is air pollution. Road transport is a major contributor to global warming, but that is not the focus of this book. Our concern is with ground level air pollution and its consequences for health. Those consequences are still the subject of some controversy as it is difficult to establish precise cause-and-effect mechanisms.

For example, the increased incidence of asthma would seem to have some relationship with increased levels of traffic, but is also influenced by the proliferation of the household dust mite in centrally heated homes with double glazing. Asthma is a debilitating and potentially fatal condition, but there are many other negative health effects that result from air pollution which, although it may be at its highest levels in 'canyon streets' in big cities, can easily affect the surrounding countryside or, in Europe, even other countries. 'Indeed, as polluted air masses drift over ... surrounding countryside, photochemical reactions lead to increased concentrations of pollutants such as ozone reaching higher levels than they do in urban areas' (Elsom, 1996, p. 7).

The health consequences of air pollution are, for understandable reasons, focused on the respiratory system. Some of these consequences may be an aggravation of already existing conditions such as asthma or chronic bronchitis, however, there are also concerns about air pollution triggering serious illnesses. 'A number of known or suspected carcinogens are detectable in vehicle emissions ... Vehicle emissions are the most important source of exposure to benzene for people in urban areas who do not smoke, are not heavily exposed to other people smoking and do not encounter high concentrations of benzene in their workplace' (Royal Commission on Environmental Pollution, 1994, p. 30).

Recent work has focused attention on particulate matter, especially the smallest particles known as PM10 which are about one-fifth of the diameter of a human hair. These are increasingly being regarded by air quality experts as more hazardous than ground level ozone and all other outdoor air pollutants because they can penetrate deep into human lungs without being captured by the natural cleansing action of the respiratory system, and collect in the tiny air sacs called alveoli where oxygen enters the bloodstream. 'Recent research indicates that episodes of high atmospheric PM concentrations correlate to increases in asthma attacks and deaths from respiratory illnesses' (House of Lords, 1996, p. 9). There is concern that longer-term exposures can lead to increased bronchial disease and an increased risk of death from heart and lung disease. A number of potentially hazardous substances have been identified in PM10. For example, elemental carbon produced during engine combustion can pick up cancer-causing chemicals such as benzo(a)pyrene and transmit them into the lung. Particulates come from a variety of sources, but diesel-powered vehicles such as trucks and buses are often the single most important source.

Although claims that air pollution has 'caused' a given number of deaths have to be treated with a measure of caution, as a particular mortality may result from a number of interacting factors (Lipfert and Wyzga, 1995), there is a growing public perception that vehicle exhausts lead to deaths that would not otherwise occur. Even if an episode of poor air quality does not result in additional identifiable deaths, it may increase the rate of hospital admissions, particularly among those suffering from respiratory diseases (Burnett et al., 1994). As citizen awareness of the health effects of air pollution increases, this creates significant pressures on public policy-makers. The central dilemma that policy-makers face is that voters want clean air but at the same time do not want to be deprived of the ability to use their motor vehicles to travel where they want when they want, any more than business wishes to have any constraints placed on its use of road haulage. Thus, statements by British deputy prime minister John Prescott that second or third cars in a household are not necessary were greeted with a measure of derision in the media. Relatively authoritarian regimes such as Singapore may have the option of taking draconian measures to curb car use, but the constraints on policy options are far greater in liberal democracies in Europe and North America.

Problems of ground level air pollution, especially in urban areas and caused principally by cars and freight vehicles, are thus becoming a focus of increasing concern for international, national and local decision-makers. There has been an increasing recognition that purely technological solutions are inadequate: there is no 'quick fix' or 'silver bullet' available that will remove the problem in the short to medium term. If policies are to be effective they need broad acceptance from affected interests and citizens in general. This is not an easy objective to achieve, but it can be facilitated by coalition-building activities and by the effective involvement of affected groups in the policy-making process. New institutional arrangements will almost certainly need to be created, not least institutions and networks that can bring together what, as Chapter 8, Marek's chapter on Berne, shows, is the wide and diverse range of actors with a stake in air quality matters. The capacity to mediate between divergent interests and perspectives is one of the preconditions of an effective air quality management policy. Political scientists thus had a capacity to contribute to the identification of effective policy solutions as much as traffic engineers.

THE CONDITIONS FOR EFFECTIVE POLICY

The central concern of this book is with effective policy formation and implementation in the area of air quality management in Western Europe and Canada. A greater emphasis is placed on the implementation of policy than on its formation. It is relatively easy to make declarations of intent to improve air quality and it is politically feasible, if costly in terms of time and effort, to pass new laws or directives on air quality or to draft implementing regulations. The really difficult task is to convert these efforts into policies that have a broad basis of support; that are feasible in the sense that they recognize implementation obstacles and means of overcoming them; and that produce outcomes rather than just outputs, in other words they lead to significant improvements in air quality.

It is important to understand how a range of governmental actors, each with their own priorities and goals, can cooperate with each other and with non-governmental actors to develop strategies and policy instruments aimed at the abatement of air pollution. It has long been recognized that 'command and control' measures are insufficient in this area of public policy and that, in the long run, more can be achieved through the more difficult but more rewarding approach of building coalitions and more or less formal networks of support (the degree of formality may reflect variations in political cultures). Whatever their precise form, innovative institutional arrangements can assist in the identification of appropriate air pollution reduction measures, the mobilization of support for such measures, and the selection of effective means of implementation and monitoring. As Giuliani notes in Chapter 3, one needs soft institutions to deal with hard problems. Building such institutions, which are bound by loose, web-like ties, rather than formal rules, is not easy.

This book examines evidence on a range of measures carried out in urban centres in a variety of countries, which involve cooperation between a range of governmental and non-governmental actors. It includes examples of policy failure as well as policy success, although in many cases the outcome is ambiguous or it is too early to evaluate the long-run effects. Through examining a number of carefully researched case studies, it should be possible to isolate some of the factors which contribute to the capacity of policy networks to develop and implement feasible and effective policies in the area of air pollution. Even if many policies are still at a relatively early stage of implementation, there may be some positive consequences. As Desideri and Lewanski note in Chapter 4, even where public policies are mainly reactive in character, they can help to legitimize the air pollution problem as a significant issue within traffic policy and facilitate a process of social learning.

POLICY NETWORKS

This book is influenced by a substantial and growing academic literature on policy networks which has in turn been influenced by the debate about 'new institutionalism' in political science (March and Olsen, 1984). As well as contributing to our knowledge of a specific and important policy area, it also represents an attempt to engage with the literature about policy networks (Coleman and Perl, 1997; Hassenteufel, 1995; Jordan and Schubert, 1992; Kenis and Schneider, 1991; for a critical view of the limitations of the approach, see Dowding, 1995).

Reference is often made in this volume to the concept of 'institutions' and some clarification of the term is necessary. Hall (1986, p. 19) uses the term institutions 'to refer to the formal rules, compliance procedures, and standard operating practices that structure the relationship between individuals in various units of the polity and economy'. Rhodes and Marsh's (1992) work emphasizes the structural relationship between political institutions as the crucial element in a policy network. While this particular approach may be too restrictive, institutions compared to networks are less flexible, more hierarchical, have a degree of permanence and a specific political mandate. By focusing on processes and interactions, the policy network approach makes the structural properties of the institutions involved in the networks more visible and comprehensible.

Policy networks can be seen as a cluster of actors connected by resource interdependencies. Exchange of resources occurs between the actors in the network, usually involving more than one resource. Resource exchanges include: information; finance; legal competences; time; consensus. Mature policy networks are characterized by a certain stability over time and shared procedural norms which govern conflict resolution procedures ('rules of the game'). Given that attempts to deal with mobile sources of urban air pollution are often relatively recent, there is a greater likelihood of encountering issue networks as defined by Rhodes and Marsh (1992, p. 14): 'Issue networks are characterised by a large number of participants with a limited degree of interdependence. Stability and continuity are at a premium, and the structure tends to be atomistic.'

Even allowing for these limitations, networks facilitate coalition building between different actors, improving the chances of the identification and implementation of effective policies. As Knoepfel following Scharpf observes (1994, p. 9), 'The network represents an important form of cooperation between market and hierarchy'. Institution building may involve mutually complementary interaction between networks and institutions. Networks may be capable of generating and winning consent for effective and feasible policy solutions, while the institutional participants in the network may possess the legal and financial resources to put the measures into effect, Marek, for example, highlights the often neglected role of the traffic police, an issue also taken up by Desideri and Lewanski in their chapter. It is important to bear in mind that some measures may be relatively cheap, or even self-financing, while others may be relatively expensive. Overall budget constraints have to be considered, including the degree of budgetary autonomy enjoyed by local and regional governments (clearly limited in the case of Italy).

For a variety of reasons, network building may be more effective in one setting than another. The chapters by Marlot and Perl on Lyon and by Marek on Berne offer a contrast between a measure of success and a measure of failure. Very specific factors may be relevant in any one case, such as the role of dominant centre-right mayors in Lyon or the contribution of neighbourhood associations in Berne. It may, however, be possible to isolate some general factors which affect the success or otherwise of a policy network.

The degree of centralization of state policy-making is one key variable. Less centralized national polities may offer more scope for a variety of local initiatives. Regional and local actors may be able to come together to pool the various resources available to them and evolve policy solutions which take account of specific local conditions. The extent to which this happens will in part be influenced by local government structures. For example, are there encompassing metropolitan authorities and regional governments (as in Toronto and Vancouver) which cover an entire 'travel to work' area? Particular attention needs to be paid to land-use planning arrangements, and the way in which they inhibit or facilitate effective air quality policies. As Larrue and Vlassopoulou point out in Chapter 6, in France for a long time air quality issues were excluded from urban mobility plans.

The significance of decentralization and its organization may be enhanced in polities where the national state is perceived to be ineffective. Despite a widely held perception that Italian cities are unable to participate in effective governance, Chapters 3, 4 and 5 in this volume suggest that they are capable of developing and putting into practice policy innovations. Indeed, the ineffectiveness of the national state provides an imperative for them to develop their own solutions to pressing policy problems. Even if some of the measures taken constitute an emergency policy style, crisis may, as Desideri and Lewanski suggest, offer opportunities for innovation and social learning. Such measures may serve as useful policy experiments and lead to long-lasting changes, as well as contributing to changes in belief systems.

While the ineffectiveness of the nation state may sometimes be a stimulus for local action, it is important not to overlook the importance of what is done at the national level. It is all too easy to accept a new conventional wisdom that states that all meaningful decision-making takes place at either a supranational or subregional level. At the very least, the nation state provides a legislative framework for what happens at regional and local level, and it is also likely to be a significant provider of financial resources. Indeed, in Canada a sceptical view might be that the main role of the federal government is to act as a mechanism for the redistribution of funds. The nation state also retains many of the resources of legitimacy in a society and will have to make decisions about how far it is prepared to share those resources with regional and local governments.

In Chapter 6, Larrue and Vlassopoulou trace the evolution of national policy in France, drawing attention to the way in which the 1996 law has opened up the transport policy network to air policy issues; and this emphasis on the influence of the central state is echoed in Marlot's chapter on Lyon. One also needs to consider the degree of cohesion and convergence of societal interests and Marlot's chapter shows considerable differences in the perspectives of central city dwellers and of the inhabitants of the suburbs. In Vancouver, in contrast, the geography of pollution gives governments in the Lower Fraser Valley, which receives pollution from the city, incentives to cooperate with governments in the Vancouver area. The Greater Vancouver Regional District provides an institutional framework which facilitates such cooperation.

Resource distributions between state and societal actors are also important. The presence of a strong 'civic culture' with representative and well organized intermediaries between governing authorities and the citizen may facilitate the design of feasible policies. As Marek's chapter on Berne shows, neighbourhood associations in Berne occupy a distinctive position as they represent the interests of both individual citizens and local commercial concerns. In Vancouver, the Board of Trade (chamber of commerce) has deliberately set out to be an inclusive organization, bringing into membership bodies such as school districts and establishing task forces on problems such as air pollution embracing a wide range of interests and perspectives (interview information).

It is also important to take into account the existing core beliefs of the various participants in an issue network. As in France, air pollution may be kept off relevant policy agendas, leading to tightly controlled policy networks which do not even accept that there is a problem which requires action. Even when pollution from mobile sources is on the policy agenda, conflicts between the belief systems of different participants may inhibit progress in policy design and implementation, or may even block the effective functioning of the network.

Advances in scientific understanding about air quality problems tend to diffuse relatively quickly among relevant professionals through a variety of mechanisms ranging from journals to conferences and, increasingly, electronic sources. Indeed, the Internet should facilitate the more rapid diffusion of new scientific knowledge and discussion of policy solutions. There is still a place, however, for the kind of informal grouping of air quality experts which meets regularly in Vancouver for the exchange of ideas. As well as scientific understanding and professional norms, one also needs to look at the stance of business on air pollution matters. In Vancouver, the perception that the deterioration in air quality could affect the important tourist industry convinced the Board of Trade that air pollution was a priority issue, but this imperative was absent in Toronto.

The substantive focus of this book is to be found in problems of urban air pollution arising from mobile sources and the effectiveness of different policy initiatives aimed at their control and abatement. The concept of the policy network is used as a central analytical tool and an attempt is made to assess which kinds of networks produce effective policy measures under which circumstances. How are these complementary analytical and substantive objectives pursued in the individual chapters?

THE CONTRIBUTION OF THE CASE STUDIES

The way in which the configuration of policy networks can affect the development and implementation of policy is emphasized in the comparative analysis of Toronto and Vancouver by Perl, Hargraft and Muxlow (Chapter 2). There were a number of significant differences between the two cities. For example, the geography of dispersal of air pollution differed considerably. In Toronto air pollution was diffused over the neighbouring region, while in Vancouver it became trapped in the Fraser Valley, leading to higher levels of pollution in neighbouring communities.

On a political level, Vancouver had an effective regional political structure charged with dealing with air pollution problems while in the Toronto area authority was more fragmented. Above all, Toronto had three policy networks with an interest in air pollution matters, one of which, the automobile policy network, enjoyed substantial economic displacement and was resistant to a number of policy solutions. Vancouver had an encompassing policy network dealing with air pollution questions, but this did not guarantee effective policy-making, particularly at the implementation stage. The structures in place in Vancouver allowed technocrats to develop effective policy solutions, but their remoteness from the citizens limited the depth of their political support. Perl, Hargraft and Muxlow conclude that an encompassing policy network is a necessary condition for an effective air quality management policy but not sufficient on its own. The most effective policy solutions are often the least politically palatable.

In Chapter 3 Giuliani complements the other Italian contributions by focusing on the regional level. He presents a careful analysis of a survey of policy makers working at different levels in the field of air pollution regulation. Since 1966 the network of agencies and actors involved in air quality management has become wider and more complex, recalling Marek's comment in Chapter 8 that bigger is not necessarily best. Giuliani also notes that 'resources of possession', such as more staff, do not seem to explain policy impact. The evidence from the analysis suggests that there is more vertical than horizontal coordination in the Italian system, although even that is imperfect.

The analysis emphasizes that policy-makers are making intrinsically political choices, a point that it is easy to lose sight of in a relatively technical area of policy such as air quality management. One of the main generalizable findings of this chapter is the importance of the link between the cognitive models which policy-makers prefer and the web of relationships they tend to activate. What policy-makers regard as the correct problem definition and appropriate policy solution varies from one region of Italy to another.

There are also significant variations in the policy models favoured by policy-makers. On the one hand, there are formal, closed and hierarchical approaches; and, on the other, an approach which is more informal, inclusive, consensual and concerned with policy-takers' reactions. In part, these different approaches would appear to be rooted in the diverse regional political cultures of Italy. Giuliani constructs, however, a more general argument that there is a clear correlation between an open and consensual attitude displayed by policy-makers, their actual involvement in wide-ranging networks with other actors, and positive policy outcomes. Building soft institutions, he argues, is a hard task but a worthwhile one.

Policy-making on air quality management in Italy is in some respects more politicized than in the other countries studied. If anything, the level of public demand for air quality management policies is lower in the other countries covered, although Desideri and Lewanski show in Chapter 4 that levels of public concern are rising. However, the implementation of policies is often delayed by turf fights, partisan rivalries and tensions between different levels of government. The superior position of organized economic interest groups in relation to the more diffuse interests of the victims of air pollution is not a specifically Italian phenomenon. However, the shopkeepers appear to be particularly effective opponents of new traffic control policies and there are no arrangements for balancing their interests with those of residents as in the Swiss neighbourhood associations. The level of passion raised by traffic questions appears to be particularly high with protests turning to violence and, in one case, an assassination of a traffic official.

Many Italian air quality policy initiatives appear to be reactive and driven by European legislation. Some of the measures being adopted, such as charging for parking, were adopted over thirty years ago in other European cities. The municipality is a focal but not monolithic actor in policy-making, but it lacks adequate financial autonomy and the necessary capabilities to develop and implement policy. Indeed, the lack of an appropriate technocracy is a serious failing throughout mobile source policies in Italy. In many ways, the Italian form of compromise is to take tough decisions – which are then not implemented. Serious implementation and enforcement deficits are common and are compounded in the case of traffic management by the fragmentation of the police and their lack of interest in traffic problems.

In the absence of clear-cut, tight and stable coalitions in urban traffic policy, Desideri and Lewanski use the metaphor of a policy web to capture the policy process. There is a resort to an emergency policy style which encourages the use of measures which are dramatic but not very effective. They suggest, however, that emergency policy measures do have one positive effect which is the promotion of a process of social learning and a change in the actors' belief systems. There is now a shared recognition that there is a problem that has to be tackled. Effective policies will require more sustained support from the centre, a stabilization of policy networks and a greater seeking to change individual attitudes and behaviour.

In their chapter on Turin (Chapter 5), Bobbio and Zeppetella contrast the failure of the introduction of a restricted traffic zone with the relative success of the introduction of parking charges, a step only recently made legally possible in Italian cities. One of the reasons why the Turin Restricted Traffic Zone (RTZ) was unsuccessful may have been that the area involved was too small to allow the expected benefits for citizens bearing its costs to materialize. Such benefits as did arise were too unequally distributed with only residents and those with permits obtaining some tangible advantage from the scheme. Even when the measure was most effectively applied, it had no appreciable effect on the level of urban air pollution, forcing the city government to halt traffic throughout the city during air pollution episodes. Moreover, the number of people exempted from the scheme and the difficulty of controlling access gates greatly reduced the expected decrease in traffic flows and congestion in the city centre. Congestion decreased inside the zone, but was exported just outside it. The problem was shifted rather than solved.

The RTZ was advanced as a policy by the green movement with a strong emphasis on ecological values and change. The introduction of parking charges was defined in a relatively closed policy network of experts as a technical tool to improve traffic conditions. Bobbio and Zeppetella suggested that the way the problem was defined had a significant impact on policy effectiveness. Urban traffic entails both pollution and congestion, but stressing the former or the latter feature may make a crucial difference. Arguments based on clean air can be perceived as ideological and an attempt to reduce urban mobility by penalizing drivers. Arguments based on congestion can be couched in terms of the aim of improving urban efficiency and the benefits are readily apparent to motorists. Policies which do not define their goals primarily in environmental terms may nevertheless have more substantial pollution reduction benefits.

The Italian chapters display a complex mix of centralization and decentralization in a polity which is undergoing a general process of political transformation. In a relatively centralized and more unchanging state such as France, it is important to understand how national policy has evolved and such an understanding is provided in Chapter 6 by Larrue and Vlassopoulou. They argue that the prevailing definition of what constitutes air pollution policy reflects the particular configuration of actors involved. In order to redefine the policy problem, it is necessary to restructure the policy system, either by changing the belief systems of existing actors or through bringing new ones into the policy network.

The chapter shows how air quality management policy in France was dominated for a long time, probably longer than in most other countries, by a preoccupation with stationary sources, with policy development strongly influenced by industrial interests. Pollution from mobile sources was dealt with in a tightly closed policy network made up of the automotive industry and the ministry of transport. Air pollution reduction was kept off many relevant policy agendas. For example, it was not an objective of urban mobility plans until recently.

In France, as in Italy, European Union directives played an important role in breaking the policy deadlock and at least stimulating some reactive policy developments. New air quality planning arrangements will lead to the enlargement of policy networks. In particular, the redefinition of the principal problem as pollution from cars will change the composition of clean air policy networks at the local level. In contrast to Italy, national developments facilitate local initiatives rather than local initiatives leading to a range of policy experiments. In more general terms, what this chapter illustrates is the interrelationship between the perception and definition of a policy problem and the composition of policy networks over time. Changes in belief systems can lead to the enlargement of policy networks, but changes in policy networks can challenge existing belief systems. In the French case, central definitions of the problem led to the retention of rather exclusive policy networks for a longer period of time than in other countries.

In contrast to Canada, Italy and Switzerland, French experience is characterized by lack of local policies on air pollution, as illustrated by the analysis of Lyon in Chapter 7 by Marlot and Perl. At an important nexus on the French road network, Lyon generates considerable amounts of local traffic, and efforts to deal with the through traffic by constructing new roads are offset by the use of those roads by local commuters. Historically local transport policy favoured the car which was associated by long-serving local mayors with modernity and prosperity.

Using Coleman and Skogstad's (1990a) terminology, the public transport network is characterized as clientele pluralism while that associated with highways is typical of pressure pluralism. The public transport network is characterized by a complex range of actors with overlapping roles and responsibilities. However, the central actor is the public transport provider which, given competition between a number of actors, has been able to use its coordinating role to gain a measure of political autonomy. The vacuum created by the dispersion of state authority and the low organizational development of societal interests is thus filled, but by an organization dominated by an engineering philosophy which often pursues costly, suboptimal solutions to problems. In contrast, the highway transport policy network is made up of strong organized societal interests which advocate policy priorities to a few state agencies with distinct interests.

The overall result is a highly disjointed set of policy networks which is unpromising for the prospects of air quality management initiatives. Unlike Vancouver, which is concerned with its reputation as a place which offers a high-quality living environment, the local elites are not motivated to prioritize air quality problems and the general public is not mobilized to push the issue up the quality agenda. At a more general level, the authors draw attention to the contrast between the globally organized automotive industry with its capacity to respond to new challenges in contrast to more locally organized public transport services where it is more difficult to learn from the experience of specific systems.

As in Italy, however, the three crises of congestion, pollution and finance may combine to force change through an emergency policy style. The new role of the environment ministry in national air pollution policy may also act as a force for change. Drawing contrasts with the Swiss experience, Merlot and Perl emphasize the value of democratization of traffic policies as a means of creating a constituency for change. However, one comes up against the problem that the most effective measures are the least popular: in Lyon, the charge for a new toll road was halved after public protests.

Chapter 8 by Marek is based on a careful analysis of the implementation of four policy measures in the Swiss city of Berne. Berne has traffic problems characteristic of many medium-sized cities, although the Swiss federal structure gives special responsibility for devising and implementing solutions to the cantonal and city level. In practice, many responsibilities fall on the city as the regional association of municipalities is afflicted by internal conflicts and lacks authority.

The implementation of the measures discussed in Berne took place against a background of political instability and increasing fiscal deficits. The measure examined in greatest detail, the introduction of a 30 kilometre per hour speed limit in certain areas, started off as a safety and noise reduction measure, but came to be seen as a contribution to the reduction of air pollution. An interesting aspect of Marek's chapter is the way in which he focuses on the role of the police in policy implementation, an aspect of the structure of authority often neglected in studies of this kind.

Marek suggests that most of the policy networks studied have core actors, for example the traffic police and the transport inspectorate, with other actors more peripherally or sporadically involved. Efforts to institutionalize networks do involve costs in terms of the delay that may arise in the implementation of measures. Marek suggests that the relationship between the structural characteristics of a policy network and policy success may be weaker than supposed with negotiations as an important intervening variable. Given the importance of negotiations, a large policy network may not be the most desirable one. There probably is an optimal size of policy network, but it will vary with the nature of the measure.

The general lessons to be drawn from these analyses will be returned to in the concluding chapter. What is evident is that the various authors found the concept of a policy network a useful tool to apply in studies covering different political circumstances and emphasizing a range of aspects of air quality management policy. The management of air quality problems arising from mobile sources is the substantive problem tackled in the book and the notion of a policy network, along with that of belief systems, is the methodological tool that is deployed.

2. When policy networks collide: the institutional dynamic of air pollution policy-making in two Canadian cities

Anthony Perl, Jane Hargraft and Kevin Muxlow

INTRODUCTION: WHAT POLICY NETWORKS CAN TELL US ABOUT AIR POLLUTION POLICY-MAKING IN CANADA

This paper compares air pollution control efforts in Canada's two largest English-speaking cities, Toronto and Vancouver. Policy-making will be analysed using the policy network perspective, which seeks to identify organized patterns of interaction in a policy community. Wright (1988, p. 606) defines policy networks as 'those actors and potential actors drawn from the policy universe who share a common identity or interest'. Coleman and Skogstad (1990a) take a more spatial perspective, characterizing the policy community as a locus of politics, the place where society's problems are contested and solutions are crafted. The ways in which policy networks share political power internally and compete or cooperate with external organizations, including other policy networks, can reveal institutionalized capacities for problem definition and instrument choice. Based on the evidence presented below, understanding the internal and external configuration of policy networks helps explain how a new initiative will be launched, but is not suffficient to predict the effectiveness of implementation, or the ultimate success of outcomes.

The comparison of air pollution control efforts in Toronto and Vancouver suggests that both optimism and caution are called for when assessing what political institutions that structure relationships among state and society actors can contribute to improving urban air quality. There is room for optimism in the finding that a good 'fit' between policy networks and policy problems will facilitate programme initiation. Vancouver's early adoption of a regional air quality management plan can be attributed to an encompassing policy community that embraced multiple levels of government and a broad cross-section of economic and societal organizations. By contrast, Toronto's adoption of air pollution control policies has been delayed by the need to reconcile multiple policy networks rooted in different political jurisdictions.

But to actually improve urban air quality, the air pollution control policies that are adopted must succeed. Here, caution is called for in extrapolating from apparently successful efforts to launch policy. Policy networks that do embrace new initiatives offer no guarantee of successful outcomes, especially in circumstances when some policy instruments provoke subsequent political controversy. Vancouver's policy implementation points to such possibilities, where a politically palatable vehicle emissions inspection and maintenance programme has delivered modest outcomes in pollution abatement in relation to the resources expended, while controversial instruments like road pricing or pollution surcharges on car use remain untested. Making the most of opportunities to implement air pollution policy may depend upon the extent of policy learning, which depends in part on policy networks' receptivity to ideas and experiences outside the network. As will be seen below, policy communities play a key role in getting air pollution initiatives off the drawing board and on to a learning curve of both technical capacity and political experience.

PROTECTING THE AIR: THE LEGISLATIVE FRAMEWORK AND INSTITUTIONAL CONTEXT

Air quality management and motor vehicle regulation fall within a typically Canadian pattern of shared jurisdiction between the federal and provincial levels of government. Following the establishment of the Canadian Environmental Protection Act in 1988, the federal government established national air quality objectives on the basis of federal–provincial negotiations (City of Toronto, 1993). These federal objectives, however, are not binding, and provincial governments may adopt federal levels as enforceable standards, or simply develop their air quality regulation consistent with the federal guidelines (Baar, 1992). Policy jurisdiction over automotive emissions is shared as well. Transport Canada, a federal ministry, establishes binding tailpipe emissions standards for new vehicles (City of Toronto, 1993). Once those vehicles are on the road, emissions regulation is left to the provinces (Anderson and Woudsma, 1996).

The jurisdictional intricacies of Canadian environmental policy do not stop with shared responsibilities between Ottawa and the provinces. Regional governments and municipalities further crowd the institutional landscape by developing and implementing environmental initiatives in Canada's urban areas. Regional governments like Metropolitan Toronto, the surrounding jurisdictions of York, Peel and Halton, as well as the Greater Vancouver

Regional District are distinctive institutions that have differentiated Canadian urban development from that of the United States (Goldberg, 1986). While urban–suburban rivalry took a toll on many American cities during the 1960s and 1970s, Canada's regional governments reduced conflicts over taxation, physical infrastructure, and land use planning by depoliticizing policy choices. Unlike the American suburbs' contest to win economic development away from city centres, and American cities' counter-efforts to recapture employment and taxes, Canadian urban areas developed policy in a more inclusive and consensual network through the early 1980s. Societal interests had less ability to play one jurisdiction off against another, and thus more incentive to reach accommodation with regional government. Perhaps the greatest drawback to Canada's cooperative policy network dynamic in urban affairs was that, by depoliticizing conflicts, regional government also insulated policies from public input.

Canada's regional governments have, until recently, kept some distance from their citizenry. Provincial governments, which hold constitutional authority over municipal affairs, intentionally circumscribed the democratic participation in regional governments when legislating them into existence (Magnusson, 1983). Given the urban concentration of Canada's population, the prospect of metropolitan politicians winning more votes than a premier was both real and unwelcome to provincial leaders. Regional governments were thus initially appointed or indirectly elected, leaving urban citizens with limited awareness of, and influence over, their activity. Only in 1997 has Metropolitan Toronto gained an elected mayor, who will have more constituents than any other Canadian officeholder. Elsewhere, urban politics fits comfortably into Canada's political dynamic of elite accommodation (Tuohy, 1992), in which a few leaders exercise the power to arbitrate societal disputes with limited public input.

Canada's multi-layered institutional context raises both the administrative skill and the political stakes involved in addressing urban air pollution. For policies to move beyond the drawing board, different governments must work together, or at least avoid discord. When there is a political will to address environmental problems, and when intergovernmental relations are managed skillfully, policy initiatives can be launched. But when politicians are reluctant to assume responsibility for difficult problems with the risk of visible failure, Canada's overlapping jurisdictions provide plenty of cover.

Harrison (1996) has noted that the intergovernmental division of policy responsibilities in Canada encourages a dynamic of 'passing the buck'. Policies that are in place also tend to suffer from undercontrol, a phenomenon where Canadian conventions of responsible parliamentary government give public offficials considerable discretionary power, including the flexibility not to act upon broadly worded guidelines or statutes intended to protect the environment. In such a heterogeneous institutional context, the number of networks engaged in policy development can play a powerful role in orienting governments' actions and inactions. The contrast between Toronto and Vancouver highlights the degree to which policy networks can facilitate urban air quality initiatives.

TORONTO: THREE DISCONNECTED POLICY NETWORKS

In the Greater Toronto Area (GTA), the extended and expanding metropolitan region surrounding Canada's largest city, the very existence of an air pollution problem has been hotly disputed. According to the Ontario Ministry of Environment and Energy, the GTA's air quality, along with that of the entire province, has been steadily improving over the last 20 years (OMOEE, 1995). Where the federal and provincial governments have exercised their regulatory authority, pollution levels of sulphur oxides, nitrogen oxides and lead have decreased dramatically. Despite this success, other pollutant levels have been increasing.

Ground-level ozone (GLO), particulate and fine particulate matter levels all rose during the same period in which other pollutants declined. Studies by Burnett et al., (1995) and Burnett, et al., (1998) identify these emissions as serious public health threats. Between 1970 and 1990, 2 per cent to 4 per cent of the deaths from heart attacks and respiratory disease in the GTA could be attributed to air pollution. A study by Campbell, et al. (1995) identifies the automobile as a major source of the GTA's air pollution. The automobile generates 93.3 per cent of the carbon monoxide, 63.4 per cent of the nitrogen oxides, and 37.5 per cent of the particulates in Toronto's airshed (OMOEE 1995, p. 79). Vehicle emissions can be reduced and new pollution control technology has brought reductions of up to 90 per cent (Deakin, 1993). However, because the number of vehicles and the length of trips have outstripped technology improvements (Atkinson et al., 1991; Mennell, 1995; City of Toronto, 1993), cleaner cars are not translating into cleaner air for Toronto. These limits to technology demonstrate the need for policy instruments that reach beyond the design capacity of the automotive sector.

Despite mounting literature which claims that the GTA's air quality produces negative effects on public health, and the evidence that industry-based solutions are not sufficient, consensus on adopting a more active approach to air quality management has been slow in coming. The provincial government has only recently announced that a vehicle inspection and maintenance programme will be phased in over five years (Rusk, 1997). This incremental approach will be administered by garages and service stations, although the precise arrangements are still being worked out. For the moment, however, disparate policy networks maintain distinct definitions of the policy issues in question. Future outcomes will depend on reconciling the viewpoints of at least three policy networks that have addressed urban air pollution in Ontario.

The first policy network may be labelled 'Environmental Professionals,' it comprises scientists, engineers, lawyers, planners and other experts who take an on going interest in environmental matters. Though actors within this network share a common intellectual focus, connections among them are weak, with limited coordination between the public and private sector participants (Interview). Actors within the environmental professionals network possess an uneven distribution of political power, with scientists enjoying less influence than process oriented civil servants, who are in turn controlled by their political masters and lack the entrepreneurial inclination to champion policy initiatives.

At the centre of this network are the provincial and federal ministries. Environment Canada and Ontario's Ministry of Environment and Energy (OMOEE) were both created only a quarter-century ago. But in Ontario, a bureaucratic predecessor dating from the late 1950s, the Ontario Water Resources Commission, brought a distinctive professional orientation to the new Ministry. Many of the top environmental bureaucrats in the OMOEE were first recruited into government to address water pollution issues. Before the mid-1980s, when Ontario began broadening its environment ministry's programme capacity, the environmental professional's network paid limited attention to air pollution (OMOEE, 1992).

The OMOEE's staff size and resources nearly doubled in the late 1980s and early 1990s, fuelled by growing public concern about the environment and supportive parties in power (Interview). More resources meant both the intensification of existing efforts, such as regulating industrial polluters, as well as addressing new policy problems like reducing urban smog. However, with rapid growth came a dilution of administrative effectiveness; the chain of command from the deputy minister to the frontline staff extended in some cases through seven tiers.

A change of government in 1990 added to the confusion. When the social democratic New Democratic Party (NDP) government took power, environmental activists who had previously clashed with the OMOEE suddenly became public offficials. These environmentalists' ambitious agenda pulled policy-making in disparate directions. For example, while the bureaucracy had one set of scientists and engineers working on air pollution issues, the minister's office had policy advisers developing parallel initiatives for both ground-level ozone and stratospheric ozone. Confusion was also manifest among government analyses and assertions. For example, while NDP politicians had asserted that poor urban air quality was affecting people's health, staff studies showed both that air pollution was improving, relative to standards, and that these standards were out of date (Interview).

Environmental professionals and politicians within Ontario's government during the early 1990s were ill prepared to forge a working relationship with a second policy network that came to focus on air pollution. Ontario's automotive policy network, as we label it, comprises industry, petrochemical companies, vehicle dealers, road construction contractors, financial institutions, organized labour, automobile associations, and a number of provincial and local government transportation agencies. Members of the automotive policy network disputed the need to regulate urban air pollution even before the environmental professionals began their work. The fact that mobile sources were known to generate the majority of pollutants in Toronto did not count for much. Instead, the automotive policy network defined the 50 per cent of Ontario's ground-level ozone precursors generated by sources in the USA as the major policy problem requiring government attention (Municipality of Metropolitan Toronto, 1996). The costs of reducing the remaining half were seen as prohibitive, since they would likely derail the province's economic engine of growth.

The auto industry's role in resisting regulatory intervention such as clean air policies is well known (Goddard, 1994; Flink, 1988; Crandall et al., 1986). Vehicle manufacturers and petrochemical companies alike have been even more strident in their opposition to economic instruments such as fuel taxes or road pricing that could be used to recover the social costs of automobility (Atkinson et al., 1991; Eck, 1991). Rounding out the opposition to an increased user-pay system are car owners themselves (Sperling, 1991). The automotive policy network is bound together by its members' opposition to regulatory activity that would concentrate the automobile's environmental costs on either consumers or producers.

The automotive network's power to constrain policy initiatives arises from the economic significance of the sector. Automotive manufacturing is the single largest sector of Ontario's manufacturing economy: manufacture, sale and service of motor vehicles account for one out of every six jobs in the province. Given this considerable economic weight, it is not surprising that both the federal and provincial governments have been carefully attuned to the policy priorities expressed by the automotive network. When the labour-backed NDP government came to power, the widespread employment in the auto, oil and road building sectors turned out to be as important as the investment and profit levels had been for its predecessors. To appear 'fair' to automotive interests, the NDP spent a great deal of time listening to their concerns, more than its counterpart in British Columbia, where the automotive sector is minuscule.

The policy network representing Ontario's automotive industry could neither be ignored, nor readily convinced to accept air pollution policy initiatives. Two years into its mandate, the NDP government implemented its first steps to control urban smog. These actions were deliberately modest, and sought to lay the groundwork for future developments. For example, the government introduced new requirements for vapour recovery devices at gasoline transfer stations, but then backed down from requiring vapour recovery technology to be installed at retail gas pumps. Oil companies argued that US regulations would require vapour recovery modules to be installed in all 1997 model-year vehicles, hence the cost of fitting gas pumps with parallel technology was excessive. Given the eight-year average age of Ontario's auto fleet, relying on in-vehicle vapour recovery technology would delay widespread deployment into the 21st century.

This incrementalist compromise between government and industry changed character when the NDP lost an election in 1995. The Progressive Conservative Party, which formed the next government, had campaigned on a platform of tax cuts and smaller government, appealing to its traditional supporters in rural and suburban locations. Fiscal and programmatic restructuring quickly changed the OMOEE from a leader into a follower in policy development (Interview). Initiatives from the Cabinet office, and even from the auto industry replaced an 'in house' agenda to try and clean the air.

The third policy network can be found in the locations where air pollution's costs are concentrated, the municipal and regional governments in the GTA. Despite this concentration of environmental burdens, GTA governments have presided over the suburban sprawl of low density housing, shopping centres and office parks that precipitated an explosion of urban mobility, mainly by single occupant vehicles. GTA governments have reacted to growing automotive traffic by building more highways, while public transport has entered a downward spiral of ridership loss, fare increases and service cuts (Perl and Pucher, 1995). By the mid-1980s, rivalry between Canada's first regional government, Metropolitan Toronto, and surrounding areas in the GTA turned debates over urban smog into zero sum trade-offs between the old centre, which stood to gain from limits to mobility and restrictions on urban sprawl, and the new periphery, which saw these policies as essential to continued growth and prosperity.

The association between auto transportation, economic growth and societal well-being appears widely shared in Ontario. Beyond the boardrooms and municipal chambers where elite members of the automotive policy community deliberate, public opinion embraces automobility as creating more good than harm overall. Harrison (1996) claims that public awareness is a prerequisite for government's environmental initiatives. Government polling revealed that while Ontarians ranked air quality relatively high as a concern, the public gave a higher priority to environmental initiatives addressing toxic chemicals, water quality and waste management. Furthermore, few respondents indicated that they would change driving and auto maintenance habits, even if this improved environmental quality. Ontarians' environmental preferences may be influenced by the region's geography which offers no barriers to the arrival of US smog and also diffuses air pollution widely across Southwestern Ontario. Such physical circumstances could create a cognitive constraint to taking action, since the causes are seen to be beyond the reach of provincial or local policy.

The environmental non-governmental organizations (ENGOs) that might have focused public concern on air pollution were paradoxically weakened by the NDP forming the government. Key environmental and health policy advocates in Ontario spent most of the NDP's mandate inside government, engaged in trench warfare with the bureaucracy, rather than mobilizing societal support for new initiatives. ENGO participants who remained outside the government lacked the political expertise to develop effective influence within the professional network (Interview). When the Progressive Conservatives gained power and most transplanted environmental advocates returned to being outsiders, Ontario's government began its own multi-stakeholder consultation with participants in all three policy networks, and others. These efforts demonstrated support for a very particular kind of air quality improvement policy – a vehicle emissions inspection programme delivered by private contractors, such as gas stations, garages and auto parts stores, on a for-profit basis. In essence, ongoing differences between automotive interests, environmental professionals and local governments were used to justify a 'go slow' approach in which government would phase in private inspections over five years.

Ontario's slow start in addressing urban air pollution suggests that, for the planned initiative to succeed, some significant changes will be needed to forge an inclusive policy network that can reconcile the political and economic rivalries between local government, environmental professionals and the province's automotive interests. Exogenous factors, such as the health effects of increasing pollution levels, may have to reach crisis levels before the disputes and discord between Ontario's air pollution policy networks can be overcome.

IN THE VANCOUVER AREA: AN ENCOMPASSING POLICY NETWORK

The Vancouver region's efforts to control urban air pollution illustrate both the potential and the limits of an encompassing policy network in delivering relief from polluted air. More than anywhere else in Canada, policy actors in British Columbia have been able to launch air pollution control policies that limit emissions in a metropolitan region. These efforts have been facilitated by an encompassing policy network, backed by an active and engaged regional government. Yet implementing key elements of the policy initiative has proved far more complicated than initially foreseen. Our overview of Vancouver's air quality management will illustrate how the allure of technically based solutions to politically sensitive problems, such as the significant share of air pollution generated by mobile sources, can tempt government into using instruments that promise low controversy. This is the case with 'AirCare', the Vancouver region's motor vehicle inspection and maintenance programme, where emission reductions have arrived more slowly than the costs of pursuing them. However, in the Canadian context of environmental buck passing where inaction and indecision predominate, launching any programme to regulate mobile source emissions must count as an achievement.

Air pollution is not a new issue to governments, industry or the general public in the Lower Mainland of British Columbia. (The term 'Lower Mainland' is the local definition of Vancouver's metropolitan region.) As early as 1949, British Columbia's provincial government delegated the power to regulate commercial and industrial air emissions to the City of Vancouver. The city established a permit system for large point sources of pollution such as paper mills and cement producers, which placed a cap on emission levels but provided no incentive for reduction. Baar (1995, p. 100) notes that 'visibility of emissions or awareness of the nuisance they create' was the guiding principle motivating the permit scheme, and that such a reactive approach 'is consistent with mopping up the most visible manifestations of neglect'. As a result, permits were issued without charge, with the primary criterion being to keep smoke, odours and other pollution impacts below the threshold of public perception, but no lower. Polluters typically negotiated for a higher quota than their current output (Interview) and unless there was public opposition, such as complaints or protests from neighbourhood associations, to the established levels, local government acquiesced.

Despite the limited scope and ambition of Vancouver's permit system, the process launched a significant pattern of interaction between business and local government on urban air pollution. Vancouver's local government began building a capacity to monitor and regulate air pollution far earlier than either provincial or federal governments in Canada. A widespread conversion to natural gas for home, office, and industrial heating needs was begun in 1956, and wood-fired boilers powering sawmills along False Creek and the Fraser River were closed during the 1960s and 1970s.

Skocpol (1985) highlights the importance of state capacity as an explanatory factor in public policy development. Policy development is often limited by government's capability; thus, taking on regulatory responsibilities for air pollution early in the postwar years gave Vancouver's government an opportunity to build experience and know-how. In the same way, Vancouver's industries also entered a policy learning curve on air quality management before counterparts elsewhere in Canada. While a part of this education took the form of learning how to minimize new regulatory burdens, these firms' engagement of government's regulatory initiative did facilitate creation of a policy network in which public and private stakeholders shared a common focus and developed new ideas and norms (Coleman and Skogstad, 1990b).

By the early 1970s, Vancouver's air pollution policy community had gained enough experience to identify air quality management as a 'common pool' problem which spread beyond the political boundaries of a single municipality. Both the city of Vancouver and its air pollution permit holders could appreciate the inequity, as well as the ineffectiveness, of confining regulatory efforts within a narrow political jurisdiction. Accordingly, the provincial government, as part of its regional government policy, amended the Pollution Control Act in 1972 to designate the Greater Vancouver Regional District (GVRD), a regional government with a much larger jurisdiction, as having responsibility for air quality monitoring and planning in Vancouver's metropolitan region (Tennant and Zirnhelt, 1973). The GVRD's activity offers the opportunity to assess the advantages and shortcomings of an interjurisdictional and multi-sectoral policy network in controlling pollution.

The technical expertise that has developed to manage air quality in Vancouver comes from technocrats working for a regional government that is more remote and less familiar to most citizens than its provincial or municipal counterparts. As a result, the capacity to analyse urban air pollution and to seek its remedy developed in a part of the public sector that lacked a direct link to the public. The GVRD is directed by mayors and councillors of 20 municipalities in the Lower Mainland, who are appointed by their respective local governments. The democratic connection between residents and the regional government is thus attenuated. When GVRD experts have put forward controversial policy options, such as road pricing (Long 1993), their political masters have proven reluctant to endorse actions that the public perceived as radical. As a result, air quality management initiatives have evolved toward technically sophisticated, yet politically palatable, programmes such as motor vehicle inspection and maintenance.

During the 1980s, regional government's monitoring and planning activities created an opportunity for introducing policies to deal with air pollution in the Lower Mainland. New information, public awareness and pressure for some action by municipal governments and provincial opposition parties converged to facilitate programme creation. The new programme adopted the goal of comprehensive airshed management and introduced regulatory measures to reduce pollution from point, mobile and area sources.

Unlike the GTA, where local geography dispersed the social and economic impacts of air pollution, geography in the Vancouver region both concentrated these impacts and raised the visibility of urban air pollution. The Coast Mountains to the north, the Cascade Mountains to the south east and the wind patterns associated with the Strait of Georgia to the west interact to inhibit the air circulation needed to disperse locally generated emissions (Bovar-Concord,

1995; Mennell, 1995; Farmar-Bowers, 1996). These mountains converge at the eastern end of the Fraser Valley, effectively confining the region's air basin and concentrating air pollution impacts in Vancouver's rural periphery. Such a concentration gave regional offficials an incentive to cooperate with the urban centre. Whereas the GTA's dispersed air pollution is no worse in outlying jurisdictions than in the centre of Toronto, Vancouver's regional rivals suffer more from poor air quality, thereby reinforcing their political motivation to pursue a policy solution.

New information was also important in focusing Vancouver's decision makers and public on air pollution as a policy problem. Collecting and disseminating data was one of the first steps that the GVRD took in pursuit of air quality management. The GVRD's initial emissions inventory showed that a total of 602 400 tonnes of pollution had entered the Vancouver region's airshed in 1985 (GVRD, 1994). This inventory also demonstrated that close to 75 per cent of air pollution was produced by mobile sources (GVRD, 1995a), a far higher level than had previously been estimated. These revelations were important catalysts for policy innovation. Industries with point source emissions already controlled under the permit system gained the justification to advocate regulation of mobile sources, which caused far more air pollution (Vancouver Board of Trade, 1991). The government was warned that any tightening of air pollution regulations that did not include mobile sources would face strong resistance.

Broadening the regulatory umbrella would give point source polluters an opportunity to shift the costs of remediation to other sources. De Spot (1994) found that in 1993, 22 per cent of all permit holders contributed 80 per cent of all permit fees. BC Hydro, the provincial electric utility, demonstrates behaviour within the policy network that typifies industry's response to this new information. Baar (1996) notes that BC Hydro sought to lead the transition to mobile source initiatives by, for example, proposing to pay for van pools, bicycle paths and buying back old cars in Vancouver as a more cost effective alternative (per tonne of emissions saved) to new control technology in its Burrard thermal generating station (BC Hydro, 1993, pp. 10–11).

A key stream of new information introduced by the GVRD was an Air Quality Index that provided Lower Mainland residents with daily updates on the level of air pollution. This index soon become a fixture of newspaper, radio and television weather reports, giving the region's population a new perspective on the state of their environment. During the late 1980s, news about local environmental deterioration in Canada's major cities coincided with reports on global environmental problems like climate change, along with new initiatives to address them such as the Brundtland Commission's advocacy of 'sustainable development' (World Commission on Environment and Development, 1987).

During the same period as these revelations about the precarious state of Vancouver's air and the environment as a whole were being publicized, the GVRD launched a consultation process on a new master plan for the region. While this input revealed general concern about the environment, air pollution from mobile sources was among the highest priority issues noted by residents of the Lower Mainland. The GVRD's summary of its new 'Liveable Region Strategic Plan' states:

> Early in the process, the public rejected a business-as-usual approach to regional growth that would spread population throughout the Fraser Valley. They rejected it because it would put development pressure on farmland, increase the distance between jobs and housing, cost too much for public services and utilities, and result in worsening air pollution from increased automobile use (GVRD, 1995b, p. 2)

Instead of business as usual, public input led the GVRD's board of directors to adopt the following vision statement for their new regional plan:

> Greater Vancouver can become the first urban region in the world to combine in one place the things to which humanity aspires on a global basis: a place where human activities enhance rather than degrade the natural environment, where the quality of the built environment approaches that of the natural setting, where the diversity of origins and religions is a source of strength rather than strife, where people control the destiny of their community, and where the basics of food, clothing, shelter, security and useful activity are accessible to all. (GVRD, 1995b, p. 2)

Such a vision would be achieved through five action plans for managing air quality, green space, drinking water quality, liquid waste processing and solid waste disposal. In each of these domains, the GVRD proposed expanding the boundaries of its previous policy responsibilities. Addressing air pollution from mobile sources was the key innovation to securing better air quality. As early as 1985, the GVRD's board had begun to urge the province to launch a new air pollution control scheme to reduce emissions from mobile sources, but the provincial government which was early in its term, and controlled by the rurally-oriented and right wing Social Credit party, refused to act.

By late 1989, the GVRD, with support from regulated industry and municipal politicians, prepared to launch a new policy from below. Regional offficials informed their provincial counterparts that they would unilaterally introduce mobile source emission regulation within the GVRD's boundary (Baar, 1996). The GVRD's threat of unilateral action was effective because it was backed by both the jurisidictional capacity and administrative competence to implement an air quality programme (Interview). Underscoring that capacity, the GVRD unveiled its Air Quality Management Plan (AQMP) in early 1990. That plan's mission statement linked intervention with cooperation, noting:

The GVRD air quality management program will work cooperatively with the community to shape regional land use and transportation, encourage clean air lifestyles, and manage emissions from human activity so as to protect human health and ecological integrity both within the region, in neighbouring jurisdictions in the Lower Fraser Valley airshed, and globally (GVRD 1994, p. 5)

Vancouver's AQMP set ambitious targets for clearing the air: emissions of sulphur and nitrogen oxides, particulates, carbon monoxide and volatile organic compounds would be cut by 50 per cent between 1985 and 2000. GVRD plans implied that emissions reductions would be achieved by using a variety of transportation demand measures (TDM) such as 'discouraging the unnecessary use of the automobile and encouraging use of ... walking, cycling, and ... public transportation' (GVRD, 1994, p. 2). Faced with mounting criticism from the opposition NDP and the prospect of action by the GVRD, the Social Credit government sought to share in the public support for environmental protection by signing on to the AQMP initiative and agreeing to a vehicle inspection and maintenance programme called AirCare (Interview). AirCare has become the most visible, and costly, component of Vancouver's air quality management programmes. It demonstrates what technical solutions to environmental problems can achieve without targeting social and economic behaviour.

In moving from formulation to implementation of an air quality management programme, policy-makers had to make key choices on the policy instruments and responsibilities for using them. As in most urban transportation problems, policy options that promised the most effective results were perceived to be politically risky, while those that were politically palatable appeared less effective. Policy-makers considered two means to reduce mobile source emissions (Interview): by employing TDMs to curb vehicle mobility, or by inspection and maintenance programmes which would cut emissions per vehicle kilometre of travel. Though more economical in reducing emissions, TDM instruments outlined in the AQMP were also quite controversial.

Provincial politicians viewed the TDM as an instrument which infringed on people's freedom to travel and would generate a backlash of public opposition (Interview). Instituting TDMs would also require government to invest heavily in revamping public transit systems in order to provide a viable auto alternative (Interview; *Vancouver Sun*, 1996). Use of TDM's characteristic pricing instruments (for example, road tolls and pollution surcharges) might trigger a tax revolt, or at least significant intragovernmental conflict over the use of those funds (Bohn, 1996). Finally, TDMs would also distribute costs unequally within the region, possibly initiating a core-periphery conflict.

On the other hand, inspection and maintenance programmes were viewed by political elites as an effective and popular regulatory pill that the public was prepared to swallow. American experience, particularly in California, appeared to demonstrate that vehicle inspections and maintenance could mitigate mobile source emissions without reducing mobility. Inspection and maintenance programmes were also politically popular, based on GVRD opinion surveys revealing that a large number of residents supported an inspection programme (Interview). Furthermore, as a publicly administered regulatory programme, AirCare would also create jobs that government could take credit for in both inspection and repair facilities. Thus, the administratively ambitious and politically palatable AirCare project was given priority over less intricate, but more authoritative, policy instruments such as road pricing or tolls on the Lower Mainland's bridge crossings.

Implementing the AirCare emissions inspection programme that assessed every vehicle in the Lower Mainland would be a complex task. The GVRD had administrative experience over regulatory programmes which were similar both in requirements and in financial magnitude to the AirCare programme (Interview). Vehicle safety inspections had originated at the municipal level in Vancouver and, within the province, the GVRD's staff had the most experience in air quality measurement and modelling. Moreover, through managing emissions reduction programmes at large incinerators and steam generators, the GVRD also had experience in the management of capital intensive projects.

However, provincial endorsement of AirCare brought provincial administrative leadership. Although the same bureaucrats from various levels of government continued to meet throughout AirCare's design and introduction, functional responsibility came to rest in the British Columbia Ministry of Highways, Motor Vehicles Branch. In principle, this was the agency of government with the final say over mobile source regulation in the Lower Mainland. In practice, it was not an agency with a great deal of experience in or knowledge of air quality management.

In meeting the new responsibility for vehicle emissions management, the British Columbia Motor Vehicles Branch sought to supplement its limited experience with outside expertise, so American consultants who had worked on inspection and maintenance programmes in California were called upon to help design AirCare. However, importing expertise raised the problem of fungibility, the appropriateness of transferring knowledge and methods that have been developed to solve one policy problem to a different context (Rose, 1993). As events proved, the American consultants' knowledge base of vehicle emissions was restricted to California experience, bringing with it limitations that would hinder the AirCare programme down (Baar, 1995). For example, one of the ways in which British Columbia differed from California that was not adequately recognized by the consultants designing AirCare can be found in the characteristics of the vehicles being regulated. California had a homogeneous fleet of vehicles to be inspected, in which all cars built in the same model-year were mandated to have the same pollution control devices. Western Canada's fleet was heterogeneous; in Canada, vehicle manufacturers would sell a mix of US specification vehicles and others that had been recalibrated to use different (and fewer) pollution control devices. Baar (1996, p. 13) finds that 80 per cent of 1984–85 model-year vehicles had had their pollution control devices recalibrated to meet the lower Canadian emissions standards. This lack of standardization in western Canada's auto fleet limited the effectiveness of an inspection programme which began with a visual check of pollution control devices to ensure that no tampering had occurred.

Because of this and other constraints arising from imported expertise, AirCare got off to a slow start in 1992, and has delivered the projected air quality improvements behind schedule and at a higher cost than was forecast. This is not to deny AirCare's positive impact. The entire AQMP, of which AirCare is only a part, has been estimated to deliver between $2.3 billion and $5.1 billion in net benefits to the Lower Mainland between 1994 and 2020 (Bovar–Concord, 1995). None the less, Bovar–Concord note in their initial assessment that 'there is a discrepancy in the timing of benefits and costs. The control costs tend to be incurred earlier in the time sequence, with major capital expenditures required in the mid 1990s when much of the control equipment is being introduced' (ARA, 1994, p. ii). Technically complex policy instruments may account for that higher cost in early years, while more cost effective but politically controversial policy measures have been deferred.

To date, the Vancouver region's AQMP has shown great promise by adopting ambitious principles. In implementation, it has also demonstrated the limitations of a technically and administratively intricate regulatory programme. In Baar's view, Vancouver's experience demonstrates that the selection of an air pollution policy 'instrument does not determine the impacts; the design of the instrument does' (Baar, 1995, p. 125). However, as one insider has observed, all levels of government have been reluctant to move toward implementing more controversial policy instruments (Interview). The Liveable Region Strategic Plan set a target of reducing 1985 air pollution levels 50 per cent by the year 2000, but the GVRD's performance forecast for the instruments that are included in its AQMP estimates only a 32 per cent reduction (GVRD, 1994). From a technical efficiency perspective, there is something to be said for selecting policy instruments which require a less exacting design process and administration than AirCare has needed. In terms of political expediency, governments may need to demonstrate that technical solutions alone cannot solve environmental problems before they propose some of the policy instruments that directly target travel behaviour.

CONCLUSION: CAN POLICY NETWORKS OFFER SPACE FOR POLICY LEARNING?

Toronto and Vancouver's air pollution policy deliberations have shown the important role that policy networks can play in either constraining or facilitating programme initiation. When a single policy network can encompass multiple political jurisdictions and economic sectors, and reach out to the larger society, launching new programmes will be easier. Vancouver's AQMP is the product of such an inclusive policy network; but, as implementation problems with AirCare demonstrate, bringing disparate actors and interests into a policy network does not guarantee effective policy.

External factors, such as geography and the role of automotive and energy industries in a region's economy, can make the path to formulating air pollution policy more or less arduous. Vancouver's inclusive policy network capitalized on geographic and economic conditions by engaging scientists, public offficials, industry and non-governmental activists in a long-term dialogue on air quality. Suffficient consensus was reached so that the results of air quality measurement could be disseminated through the media. In turn, supportive public opinion in Vancouver reduced the risk for policymakers, while inconsistent public opinion in the GTA raised the risk for decision-makers there. Finally, timing was important for the initiative's success in Vancouver, with the provincial government seeking a 'green' policy issue at the same time that the GVRD had launched its AQMP.

Despite these different external inputs, policy network characteristics do explain part of the variation in successful air pollution policy formulation in Vancouver versus the GTA. Variation in the degree of internal consensus among policy actors, the power that the various policy networks could deploy in pursuit of their agendas and the state's ability to support policy entrepreneurship all contributed to successful policy development in Vancouver and a slow start in the GTA. We deal with these in turn.

Compared to the three policy communities in the GTA, Vancouver's policy community demonstrated a high degree of integration. Internal consensus was established around a set of shared ideas regarding the significance of the policy problem and its cause. Such shared ideas complemented the participants' different interests and created a foundation for cooperative behaviour. The resulting alliance between Vancouver industries and the regional government to target mobile sources of air pollution allowed industry to shift remediation costs, while at the same time the GVRD could act to reduce social costs. In the GTA, however, neither shared ideas nor interests existed among the three disconnected, and often adversarial, policy networks. For example, with some participants defining air pollution as a transboundary issue and others seeing it as a automotive issue, policy networks remained divided on the cause of the problem. Moreover, the sweep of inclusiveness was due, in part, to the macro-institutional framework within which these policy communities functioned. In Vancouver, institutions such as the GVRD promoted a dynamic of cooperation and compromise which was legitimized by the provincial government. In the

GTA, on the other hand, the presence of several competing governments produced a dynamic of regional rivalry which was also legitimized by the provincial government.

Another key difference between Toronto and Vancouver was that the latter's policy network had reached a critical mass where a broad-based, long-term engagement among policy participants taught them how to work collaboratively, and demonstrated the benefits of such behaviour even in terms of more particular interests. Within this dynamic, participant linkages were numerous, bridging the interest differences across the network. Such connections were evident among technical experts, industry, the public and the state. In the GTA, where such a critical mass has not yet occurred, the automotive sector remains its own policy network, and a dominant political force. What's good for General Motors is implicitly accepted as being good for Ontario.

Finally, successful policy formulation in Vancouver was a consequence of an active state supporting the work of policy entrepreneurs both within and outside government. Atkinson and Coleman (1989) have observed that, for an anticipatory policy like Vancouver's AQMP to be realized, the state must be able to coordinate policy network participants and support to their entrepreneurial activities. The GVRD possessed both the incentive and the authority to play a leadership role within the policy network that fostered the AQMP's development and was therefore able to set the agenda, foster inter-governmental and societal partnerships, and maintain effective pressure on the provincial government to implement the AQMP. No such effective policy entrepreneurship occurred in the GTA; instead, the three policy networks continued to talk past each other.

By this point, it has become clear that network characteristics influenced the course of pollution policy development in both Vancouver and the GTA. This analysis has revealed that, where an encompassing policy network evolved, policy initiation occurred and network participants began the learning curve of implementation. But where a number of disjointed policy networks remain in place, policy development remains challenging in its own right. This conclusion raises the question of policy actors' capacity to forge an encompassing policy network from a fragmented set of interests. Given its structural authority, to what extent can the state forge links between disparate policy networks?

Looking elsewhere in transportation, Dunn and Perl (1996) have identified an inventory of societal partnerships that state actors can cultivate to build a political infrastructure that can support innovations like high speed rail travel. Though such political engineering can create new horizontal links between disparate communities and support policy entrepreneurship, Majone (1989) points out that institutional tinkering by the state to shift the balance of power within existing networks (for example, by choosing to share decision-making power with particular organizations) will alter the comparative advantage held by certain interests, thus raising the possibility of negative consequences, such as loss of electoral support, for visible public offficials. It is our suggestion, therefore, that only those offficials who can either live with the threat of a political backlash for unsuccessful outcomes, or can craft a strategy to avoid such blame (Weaver, 1986), will assume the role of institutional architect.

In conclusion, though an encompassing policy network may well be necessary to formulate an innovative air pollution policy, it is not necessarily suffficient to ensure effective outcomes. In order to achieve the level of air pollution abatement needed for long run urban sustainability, regulatory and/or pricing measures that exceed the current achievements of even Canada's most accomplished policy network are likely to be required. Difficult choices will have to be made regarding effective, but controversial, policy instruments, with more significant distributional impacts than current air quality management tools have produced. The key ingredient which may aid in the adoption of such future measures is policy learning. Despite all of its intellectual pitfalls (Bennett and Howlett, 1992), policy learning offers a practical opportunity to build upon current efforts and enhance the cognitive capacity designed to better link policy instruments to policy goals. In a decentralized federation like Canada, such learning is likely to be local in character.

As they work through air quality initiatives by trial and error, politicians, industries and the general public will need time to discover that politically innocuous policy instruments will not be sufficient to clear the air. As with most learning curves, the earlier that individuals and communities begin their experience with air quality management efforts, the earlier they will gain enough experience to make effective choices. In this race between collective policy-making capacity and the 'green wall' of limits to urban sustainability, the existence of a policy network that can move disparate actors on to the learning curve may make an important difference.

3. 'Soft' institutions for hard problems: instituting air pollution policies in three Italian regions
Marco Giuliani

INTRODUCTION: INSTITUTIONS AND NETWORKS

The revival of the institutionalist perspective in political science has given back to 'structures' a prominent role in the determination of the contents and dynamics of policy-making. This renewed attention should not, however, be confused with an exclusive focus on formal architectures of power and competences. In fact, the definitions proposed by distinguished new institutionalists are somehow reassuring in this regard. March and Olsen's books have clearly set comfortable boundaries to institutions, referring their analysis 'not only to legislatures, executives, and judiciaries but also to systems of law, social organization, and identities or roles' (1995, p. 27). Institutions are 'collections of interrelated rules and routines that define appropriate actions, [...are] agents in the construction of political interests and beliefs' (1989, pp. 160, 165).

Inside these boundaries, there is plenty of space for very different approaches: for traditional political scientists like Weaver and Rockman (1993), and their concept of 'governmental capabilities'; for economists like North (1990), who talks about formal and informal institutions, regulations and conventions; for scholars of political economy like Hall (1986) who defines institutions as 'rules, compliance procedures and standard operating practices'. Historical institutionalists with their interest in institutions as 'shapers of preferences' (Steinmo, Thelen and Longstreth, 1992), and rationalists attracted by the ordering role of institutions and ideas (Garrett and Weingast, 1993) are probably at opposite poles, but still under the same roof.

Each social science has its own 'new institutionalism' (Goodin, 1996), and political science itself has more than one (Hall and Taylor, 1996; Kato, 1996; Peters, 1996). For policy scholars, the stabilizing effect displayed by institutions – widely conceived – can be assured by different factors: by 'hard' institutions like the electoral system, the government structure or the legislative-executive design, as well as by 'smoother' variables such as sectoral webs linking policy actors characterized by an intersubjective understanding of the desirable goals and feasible means of the policy process. We will use here the term 'soft institutions' to address this second class of variables, acknowledging the independent role exercised upon decision-making by policy networks and cognitive variables.[1] In this sense, we won't contrast on a theoretical ground 'policy community realism vs new institutionalist ambiguity' (Jordan, 1990), since both concepts belong to the same *genus*, though to different *species*. On the contrary, we will test their explicative potential empirically, analysing a specific policy sector. To be more precise, we will compare the role exercised by organizational resources – a traditional element of a particular type of 'hard institution' with that displayed by cognitive frames and belief systems of networked policy-makers in determining the local effectiveness of urban air quality regulations in Italy.

Using both primary and secondary data, we will argue that soft elements of the policy process account for a greater amount of variation in the implementation of air pollution policies than agencies' resource endowment. The availability of money, staff, norms and information sets only a minimal threshold to their success. On the contrary, there is a clear correlation between an open and consensual attitude displayed by policy-makers, their actual involvement in wide, regular and multifaceted networks of contacts with other actors (experts, environmentalists, and so on), and the positive outcomes achieved in the investigated sector.

There are good empirical and theoretical reasons to concentrate one's attention on this more slippery side of the policy process. First of all, regulatory studies have frequently pointed at the usefulness of reaching the goals normally obtained through standard setting or economic incentives using alternative instruments, though they have often been conceived as a kind of residual strategy (Dente, 1995). These soft regulation devices include the autoregulative bargaining between actors, the cooperative definition of problems and solution, the consensual search for innovative strategies, the diffusion of information and a participatory approach to decision-making. These instruments have been explored in order to bypass the rigidities of top-down enforcement, acknowledging the obstacles stemming from different interpretations of the regulatory problem in the typical complex situation of joint action.

On a more theoretical level, most policy scholars have realized the inadequacy of simple means–ends models of decision-making, in which actors behave as computational machines inside a system of constraints and opportunities given by rigid structural factors. Policy-makers began to be considered as active subjects (Lundquist, 1980), performers of roles which they contribute to create. Certainly not selfish monads in a political vacuum, but actors whose ideas, cognitive frames, problem definitions, subjective dispositions, argumentative capacities, networking skills, symbolic appeals and aggregating abilities have some role in the policy process (Fischer and Forester, 1993; Faure, Pollet and Warin, 1995.) They contribute to the establishment of the categories for the interpretation of a normally contested political and policy reality, to the establishment of the frames that permit (or inhibit) trust and cooperation, to the realization of a context for intersubjective recognition and for intertemporal preference calculation: in short, they participate in constructing 'soft institutions'.

In order to contrast the effects of hard organizational factors and soft cognitive and networking elements, we will try to operationalize all these categories. After specifying the reasons for testing our hypothesis at the regional level, we will suggest comparable indexes of policy performance and resource endowment. The absence of a clear relation between these two measures, turned us towards the investigation of the attitudes exhibited by policy-makers at the local level. Thence, using a survey expressly realized in three different regions, we will try to picture how they define the problem of urban air pollution and portray alternative policy solutions. At the same time, we will verify how these core policy-makers relate themselves to other actors, stakeholders and policy-takers. Finally, we will outline a sort of 'geography' of these cognitive and networking elements, matching it with the odd success exhibited by the investigated regions in implementing effective urban air pollution policies.

THE POINT OF DEPARTURE: MAPS OF ENVIRONMENTALISM

Even those who have only a vague idea of Italy know that it is a nation which is highly differentiated internally. Leaving aside all the stereotypes regarding Italian society, there is no doubt that its territory represents a fascinating combination of manifold local realities; regions are administrative entities which reflect, more or less, the historical, political and cultural traditions of each territory. It would not be difficult to trace even intraregional differences – among provinces, municipalities and so on – but it all depends on the preferred level of analysis, on the type of lenses you intend to wear.

To justify the choice of regions as the most appropriate point of departure, it suffices to say that, since the 1970s, environmental policies have been formulated and implemented mostly at this level. Regional governments enjoy consistent degrees of freedom inside the common framework set up by the national legislation, and they act with the support of municipalities, provinces and other specialized local units. Moreover, most of the judiciary policy controversies are adjudicated by administrative courts established at the regional level. Apart from being the easiest level to manage in order to give an overall picture of the performance of environmental regulations, regions are thus located at the strategic juncture between administrative and political responsibilities, between national directions and street level administration. In this context, the fact that regional administrations exhibit a manifest variance in the way they manage common environmental policies is all the more interesting: it represents the typical comparative setting in which a similar impulse – the national legislative framework – produces very different outcomes. In other words, regions are definitely the most natural and fruitful points of observation for investigating environmental policy dynamics.

This is true not only for the regional government itself, but even for regions conceived as units of analysis, as geographical containers of complex policy phenomena, with their problems, resources, emergencies, cultures, citizens, firms, experts, environmental associations, and the rest. Although it is generally acknowledged that each single municipality has its own peculiar character, still the regional level of analysis holds even from a bottom-up perspective: regional loyalties and sense of belonging have increased recently; parties and environmental groups are organized locally and often display regional specificities; the socioeconomic frame is clearly differentiated.

From the perspective of decision-making, we have already anticipated that regions do not exhibit the same capacity to administer policy problems. For some scholars, this kind of ability, as well as the efficiency and effectiveness displayed, is the consequence of a long-term legacy which has nothing to do with the investigated sector (Putnam, 1993). This approach is probably useful in falsifying the more simplistic hypotheses, such as those based on socioeconomic indicators or on the level of autonomy given to local governments, but it fails to recognize issue-specific differences.

In fact, regions do exhibit huge variations even inside the specific arena of ecoregulation, as it is attested by the official reports (Istat, 1993; Ministero dell'Ambiente 1989, 1992), as well as by the continuous monitoring of environmental associations (for example, LegAmbiente 1995, 1996, 1997). Whereas it has been demonstrated that regions share a similar *political* attitude towards the environment – strictly adhering to an electoral-environmental cycle[2] – they sharply differentiate themselves at the *policy* level. Regions formulate policies that differ in quantity and quality, and they implement them to very different degrees. Using different types of data, it is possible to create a kind of typology of the environmental concern exhibited by the 20 Italian regional governments (see Table 3.1). This classification is based on the activism manifested in environmental law-making on one side, and on the implementation of those measures on the other.[3] Each region has been placed in a distinct class, depending upon its performance compared to the national average.

We can thus recognize different profiles of environmentalism: 'active' regions like Emilia-Romagna and Trentino Alto Adige, contrasted by 'passive' ones like Sicilia and Campania. But we have even more complex cases: like

Table 3.1   A typology of regional environmental performance

	Formulation +	Formulation −
Implementation −	Pugila Sicilia Campania Sardegna Molise Calabria Basilicata	Lazio
Implementation +	Marche Val d'Aosta Friuli Umbria Toscana	Trentino Alto Adige Emilia Romagna Veneto Piemonte Abruzzo Liguria Lombardia

Source: Classified upon indexes presented in Giuliani, (1997).

Lazio, in which the law-making activity largely exceeds its actual commitment in the implementation phase, raising the suspicion of a merely symbolic contribution; or regions like Umbria and Toscana, which are reluctant to introduce complex normative frames, but which can claim a positive implementation record. Beyond suggesting a variegated environmental concern – which is only partially explained by the variations in the amount of human pressure upon an area – these disparities require some kind of interpretation. The most common independent variables, such as the distribution of socioeconomic indicators, the degree of civicness, the partisan composition of local assemblies or the governmental stability exhibited by regional administrations, do not seem to be strictly connected with our categories. Eccentric positions to these kinds of hypotheses are more than simple exceptions, and the indexes of environmental attention which underlie our typology remain mostly unexplained.

In order to take a closer look at this problem, and to investigate it further from a different perspective, it was necessary to abandon our cumulative indexes and adopt a more policy-specific approach. Hence, we chose to focus on the analysis of urban air pollution policies, first concentrating on the available resources, outputs produced and outcomes obtained in each region, and then introducing soft elements such as networks and cognitive variables.

THE FIRST STEP: RESOURCES AND OUTCOMES

As many analysts have observed (for example, Liberatore and Lewanski, 1990; Lewanski, 1997), Italian environmental policy began in 1966 with an act regarding air pollution. Since then, with a policy-making style that is widely recognized as being mainly reactive, emergency oriented and of the 'garbage can' type, more than a hundred national acts – standards, decrees, ordinances and so on – have been approved on the topic. The range of offices and administrations with different tasks and responsibilities, which have been summoned over the years is impressive: to quote only the most relevant ones, we should recall the Minister, a central committee against air pollution (and many regional ones), the national Environmental Agency (also many at regional level), regions, provinces, municipalities, laboratories of the National Health Service, fire brigades and so on and so forth.

We will not discuss here the internal evolution of goals and competencies of this sector, nor the political, administrative and technical quarrels that have characterized the policy area in the last three decades (but see Signorino, 1996; Lizzi, 1997). It is sufficient to recognize that, for many reasons, the involved network of agencies and actors has constantly become wider and more complex, including all possible levels of government participation.

Given the insufficient explanatory potential of the hypotheses based on political, economical or cultural factors, a classical interpretation of the geographical variation of policy performance points at the amount of resources available to local administrations. In the literature on implementation deficits, the shortage of resources has been often considered a major cause of policy failure: poor resources, broadly defined, means poor implementation. If the shortage of funding, that is of economic supplies, is constantly quoted as the most serious of the administrative problems, being unanimously blamed also by the respondents of our survey, even the lack of other kinds of resources, such as personnel, information and instruments, worries policy-makers.

The first step of our investigation therefore consisted in verifying the correlation between the stock of organizational resources locally available, and the performance in controlling the air quality. The independent variable – resources – has been broken down into four constitutive elements, which were considered as equally important: money, staff, standards and information.[4] The dependent variable – policy impact – has been operationalized through disparate indicators representing the evolution of emissions, the measure of air quality and the use of environmentally friendly transport systems.[5] In both cases, the adoption of multiple indicators and the absence of exogenously imposed thresholds, were aimed at extending the variety of information used and at minimizing the bias due to possible eccentric values (King, Keohane and Verba, 1994).

If the original organizational hypothesis that more resources means better implementation was correct, we would expect to find a consistent positive correlation between the values exhibited by each region on our two indexes. 'Unfortunately' this has not been the case, as it is clearly displayed in Figure 3.1. The scattergram leaves no way out; there is no positive correlation between the amount of resources locally available and the performance measured on the outcome index. On the contrary, regions seem almost to align themselves along the secondary bisector, from the upper left corner to the lower right one, as it is demonstrated by the low, but negative, correlation coefficient (–0.19).[6] Some of them, for example Emilia-Romagna, seem to accomplish a lot, in spite of the comparatively low level of resources, whereas some others, such as Abruzzo, exhibit poor outcomes notwithstanding the abundancy of their organizational endowment.

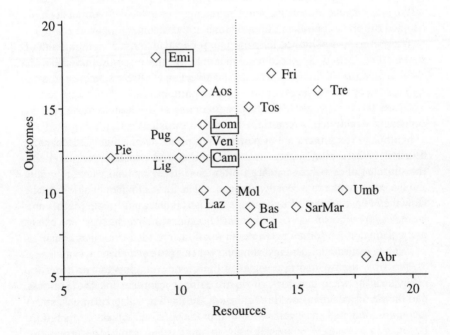

Source: Own elaboration of various data; see Giuliani (1997)

Figure 3.1 Resources and outcomes of air pollution policies

Moreover, if we try to calculate separately the contribution made by each type of resource, we will find that those that play a major (positive) role are the ones we could call 'relational resources' (rules and information), and not 'resources of possession' (such as money and staff). Once more, these findings, which contradict the public opinion and the complaining image continuously adopted by local administrators, bring us back to our original problem: neither macro variables, nor hard organizational factors can account for the variation locally displayed in the effectiveness of air pollution policies.

THE SECOND STEP: IDEAS AND NETWORKS

Having disposed of every possible theoretical misunderstanding regarding the use of the concept of institution, and disproved the most traditional empirical explanations, we have to introduce the *pars construers* of our analysis. To recapitulate: we are investigating a policy area displaying some peculiar features; the issue domain is animated by policy actors which contributed to its definition in the different stages of the decision-making process – problem outlining, agenda setting, instrument choice and policy implementation; local governments have adopted different devices in order to tackle the problem.

The preferences exhibited for particular policy solutions are not the result of sterile computations, the outcome of maximizing algorithms. Rather, these are choices which are intrinsically political, the reflection of different decision-making styles, of cooperative versus confrontational attitudes towards external actors. They are the consequence of the feedbacks of past policies, an exercise of path-dependent search into inventories of accepted measures.

In other words, the choice of policy instruments is not a purely mechanical decision. What policy-makers regard as the correct problem definition and appropriate policy solution varies from region to region, and we believe it is associated with their different performances. The ideas of policy-makers matter. How they construct causal links, and establish connections between goals and means is not without importance, not least because it affects the types and extent of the formal and informal networks of relationships they tend to establish.

These two elements, ideas and networks (or cognitive and relational variables), are probably the two most powerful and interesting challenges ever faced by policy analysts. Many scholars believe that these concepts are crucial to grasping the complexity of policy-making processes, the analytical link between context reconstruction and generalization, micro analysis and causal explanation. Unfortunately, as in every 'loosely coupled' model, these same concepts oppose every standardization effort and, above all, seem to resist operationalization. Ideas, in particular, can be communicated, sympathetically understood, or extensively explained, but not synthesized. That means that cognitive and relational variables hardly fit into research designs which aim at something more – if possible – than the in-depth case study's context reconstruction.

Our analysis tries to bridge the gap between deterministic models and postmodernist metaphors (Radaelli, 1995), exploring an explicit operationalization of both variables, and applying non-sophisticated statistical techniques to the answers given to a survey addressed to institutional policy-makers working at different levels in the field of air pollution regulation.[7] For practical reasons, the survey could not be fully replicated over the whole country in the same detail. Whereas every regional government was contacted, the provincial and communal levels were taken into account only in three regions which represented a reliable sample of local realities: Lombardia, Emilia-Romagna and Campania. These are all important regions, located respectively in the north, centre and south of the country, which have different political traditions, and have even followed divergent industrialization paths. In spite of the many differences (confirmed by the data already presented in Figure 3.1 and Table 3.1), Milano, Bologna and Napoli, their main towns, share a common concern: how to tackle urban air pollution. Given the seriousness of the problem, since the early 1990s, all these cities were included in the small number of urban areas which deserved special attention, and which were regulated by *ad hoc* governmental decrees. At the same time, especially during winter, Milano, Bologna and Napoli often gained the attention of the media for having repeatedly exceeded air pollution thresholds, and for having been forced to adopt emergency measures (mainly traffic and heating restrictions).[8] In short, the selected regions (together, perhaps, with Piemonte, Toscana and Lazio), are among the areas in which the air pollution problem is at its worst, and for many reasons they appear as worthy and unbiased case studies for an in-depth analysis.

The survey was designed to investigate the attitudes towards the problem of air pollution (and its treatment) exhibited by institutional policy-makers working in the regional and provincial governments, as well as in the main municipalities of the regions studied. How did they define the problem itself? Which did they consider to be its main features? Why was the situation so problematic? Which types of solutions did they value as the most effective and reliable ones? This investigation was carried out by asking each respondent to express his or her degree of agreement with sets of sentences or propositions. As we will specify further, the answers obtained have been submitted to factor analysis in order to extract the underlying cognitive frames.

At the same time, respondents were asked to furnish new insights on the basic features of the policy relationships they maintained: mainly how they were linked to other administrations at different levels, and to non-institutional actors such as environmental associations, external experts, businessmen, technical offices of the National Health Service, political parties and citizens. Notably, for each of them, they were asked about the frequency of meetings, the kind of problems dealt with, and who normally initiated those contacts. This has made it possible to ascertain the different degrees of centrality of the various actors inside the network, its extent, the level of complexity of the decisions taken, and so on. Although lacking the precision of a formal network analysis, which would need a complete matrix of the interactions, the survey nevertheless ensures the recognition of the most relevant features of the activated policy communities.

The survey which was undertaken during 1995, was addressed to 70 holders of 'policy positions' in regional, provincial and municipal governments, and to 20 environmental associations. Taking into account only the answers of institutional policy-makers, the average rate of return was around 50 per cent (almost 60 per cent in the case of Lombardia, Emilia-Romagna and Campania). On the whole, the ratio was lower for the regional level (only 30 per cent), and somewhat higher for the provincial and communal levels (48 per cent and 64 per cent respectively). Including the answers of environmentalists, our return sample included cases from all over the country, though not uniformly distributed: 40 per cent were from the northern regions, and the same percentage from central Italy, whereas only 20 per cent came from the south. Even so, the rates of return of Lombardia, Emilia-Romagna and Campania, our regional sample, were roughly similar, varying from 55 per cent to 58 per cent.[9]

We would like to begin this part of our analysis with a simple comprehensive indicator of connection among potential actors, without making any geographical or administrative distinction. Which are the main 'partners' of our institutional policy-makers in processing environmental problems? Figure 3.2 establishes two different thresholds in this regard, depending upon the frequency of meetings with other policy actors reported by our respondents. The diagram represents the percentage of respondents which claimed to discuss air pollution questions with the categories listed on the left: either at least once in a month; or at least once in a week.

This indicator provides only a rough guide; it cannot be assumed that the role played by municipalities is half as relevant as that played by citizens simply because the latter are mentioned twice as often. None the less, it is worth noting that the Environmental Ministry is rarely quoted and, at the most, its officials are contacted no more than once a month. Even more striking, in a supposed party government system, is the absence of political parties (Greens included) from the environmental policy scene. The administrative division of labour seem to assign a central role to provinces, with the crucial help of the local units of the Health System in charge of the continuous monitoring and evaluation of the air quality. In addition, though this is not represented in the diagram, most of the relationships activated are of the vertical type (for example, region–province, or province–commune), whereas horizontal coordination between neighbours (that is between offices belonging to administrations placed at the same level) is unusual.

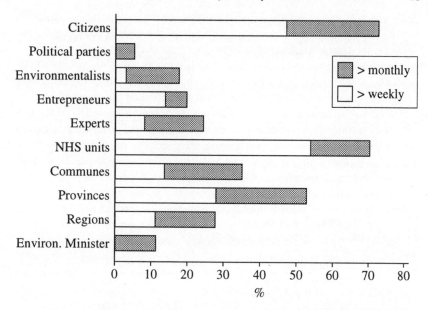

Source: Own survey (Giuliani, 1997)

Figure 3.2 Mobilization of actors with regard to air pollution problems

Leaving aside the apex reached by the contacts with the public (mostly due to emergency situations or complaints), the importance of other 'external' actors seems to be evenly balanced with businessmen, environmentalists and environmental experts being equally represented. However, the first two types of actors normally make an approach to institutions, whereas the third one is contacted by them. In no way, should Figure 3.2 be considered as a representation of the degree of centrality of the different actors in the policy network; rather, it is an indicator of the level of 'mobilization' of each policy category around the pollution issue.

It would be possible to go further in this direction by pointing to the various degrees of complexity manifested by each connection, underlining the most formal or informal relationships. We will briefly return to this point later, after having introduced our geographical discriminant. What needs to be addressed first, is the problem of the 'cognitive frames' adopted by policy-makers. It is of particular interest to go into their problem definition, looking for the way in which they conceive the origins of the policy issue itself, and the difficulties they have in dealing with it. Our respondents have expressed their opinions regarding eighteen potential obstacles to the resolution of air pollution problems, which have then been summarized in three different 'positions' through a varimax rotated factor analysis.[10] The major factor loadings are represented in Table 3.2. The higher the value shown, the more important is the statement on the left in the definition of the characteristics of the factors extracted.

Table 3.2 Cognitive maps of policy-makers regarding problem definition (factor loadings)

	Administration	Management	Citizenship
Lack of technical capabilities	0.71		–0.38
Admininstrative complexity	0.69		
Juridicial definition of the measures	0.63		
Vertical dispersion of competences	0.51	–0.51	
Horizontal fragmentation of the tasks	0.50	–0.54	
Overall indifference towards the problem	0.49		–0.36
Economical efficiency	0.48	0.59	
Lack of public information	0.38		0.65
Opposition of private economic interests		0.67	
Prevailing private well-being habits		0.62	
Search of the technical optimal solution		0.60	
Excess of private transport		0.36	
Environmentalists' rigidity		–0.62	
Absence of non-polluting transports			0.66
Opposition of many citizens			0.49
Insensibility towards public goods			0.31
Interference of political parties			–0.39
Lack of economic resources			–0.56

Source: Own survey (Giuliani 1997)

The answers tended to converge around well identified conceptions of the policy problem. We have attributed different labels to the three factors.[11] The first, which accounts for the majority of the original variety of answers, has been called 'administration' because it is mostly correlated with definitions of the policy problem which tend to stress the bureaucratic side of decision-making. The inability to reduce air pollution largely arises from organizational inefficiencies; lack of professionalism, administrative complexity, horizontal and vertical fragmentation, and absence of clear norms are the main elements considered to be the primary causes of possible implementation deficits. It would be possible to argue that this first conception partially reflects the hypothesis tested in the first part of the paper: that outcomes depend upon the stock of resources available to local administrators.

We have named the second factor 'management', because it tends to emphasize problems of optimization: the pursuit of economic efficiency and the search for the superior technical option dominate this cognitive policy model. At the same time, strong, private and well organized interests tend to oppose any expensive environmental solution. Following this interpretation, policy failures do not depend upon the administrative architecture (notice the negative factor loadings), but arise from some kind of regulatory capture, or unbalanced confrontation between concentrated and dispersed interests. Unfortunately, a positive-sum solution, which could reconcile these opposites, doesn't appear technically feasible. On a day-to-day basis, materialist values (such as the use of one's private car) still defeat any post-materialist alternative. Traditional habits and private short term well-being are still unsurpassed, even in the proxy cost–benefit calculations made by the population.

The third factor has been labelled 'citizenship', although the term 'public interest' would have suited equally well. The statements most loaded on this factor reflect a common concern towards consensual decision-making practices, and public goods problems. The public have to be kept informed, in order to acquire their consent and to encourage environmentally beneficial habits; their opposition – that is the sum of thousands of selfish non-compliant behaviours – is one of the major causes of policy failure. Whereas the search for new, technically advanced, sustainable solutions has to continue, it is not in the political, administrative or organizational field that we have to look for innovative answers to the air pollution problem, but in the construction (or reconstruction) of a 'cooperative capital'. First of all, we have to remember that air is a public good. Its public character calls for *collectivized* decisions which may possibly, even have to be *collective* decisions.[12] People are not naturally indifferent to environmental problems because of their inherent nature, and their free-riding behaviour tends to persist only while they remain uninformed or excluded from any possible influence.

We could speculate about the potential power relationships which underlie these different forms of cognitive framing of the policy issue. The first one, the 'bureaucratic' image, could be associated with strictly top-down relationships aimed at solving coordination problems between public administrations. There is no room for outer contributors, either in the simple form of external consultants, or in the more complex one of the internalization of the policy-takers' points of views. These extremely isolated relationships are chiefly maintained to fulfil legal obligations, or to accomplish such bureaucratic tasks as authorizations, controls and so on.

The second, 'managerial', problem definition could suggest the establishment of policy networks which include actors normally outside the traditional politico-bureaucratic circuit. Beyond those institutional policy-makers, the contribution of technical specialists legitimized by their expertise, and of interest groups' representatives sustained by their organizational strength, could be positively valued. One of the major resources exchanged within these networks, as well as the implicit prerequisite for its access, would be the control over information which is crucial for the technical or economic optimization which is at the core of this policy frame.

Finally, the 'consensus-building' attitude which underlies the third factor extracted by our analysis suggests the inclusion of a more varied range of non-institutional policy actors. These networks should be more extensive and far-reaching than the preceding ones, including the most interested policy-takers in an enlightening definition of the policy goals (Weiss, 1979). Trust, rather than authority or information, should be what binds these types of relationships, suggesting the establishment of policy communities rather than unstable issue networks.

THE THIRD STEP: A SIMPLE GEOGRAPHY

In order to find empirical support for our hypothesis that cognitive frames are significantly correlated to formulation dynamics and implementation outcomes, it is necessary to demonstrate that the dominant problem definition given by policy-makers operating in the regions investigated is not the same. In other words, that the three factors extracted, together with their respective stylized networks, should characterize differently the attitudes and perceptions of the respondents, depending upon their geographical location. If, on the contrary, cognitive maps did not exhibit that uneven distribution, it should be argued that they are irrelevant. By calculating the scores attributed on each factor for all the policy-makers and then computing the regional average, we immediately obtain an answer to this research question. Figure 3.3 compares the scores obtained by respondents of Lombardia, Emilia-Romagna and Campania. The higher the values for each factor, the more the actors' cognitive maps and their definition of the air pollution problem are shaped by the constitutive elements of the underlying dimension.

The simple fact that factor scores are so unevenly distributed appears to be a first corroboration of the proposed argument; but there is more to it than that. First, the two regions whose policy-makers present the most clear-cut configurations of attitudes – Emilia-Romagna and Campania – are just those ones (in our sample) neatly opposed regarding their policy performance. In Lombardia the cognitive profiles of policy actors seem more uncertain, presenting low values on each dimension, as if they were composed by a mix of contrasting elements. Not surprisingly, this region scored exactly in the middle in the outcome index, thus further confirming the correlation between 'policy ideas' and 'policy effectiveness'. Second, but even more important, the frames that emerge from the analysis of the actors' definition of the policy problem appear to be confirmed by the types of solutions preferred and implemented, and by the policy-making

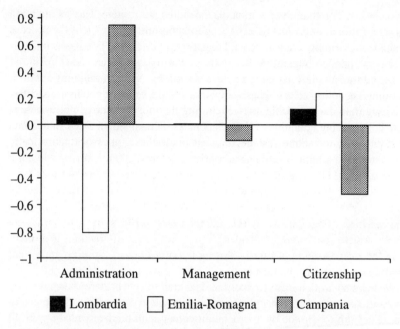

Source: Own survey (Giuliani, 1997)

Figure 3.3 The geography of cognitive maps

style adopted by institutional decision-makers. Let us now take a closer look at the distribution of the factor scores in Figure 3.3.

In Campania, policy-makers tend to attribute possible shortcomings in the formulation and implementation of the air pollution policy to administrative problems, that is to difficulties due to the lack of technical resources, to organizational complexity and to unclear norms. It is the traditional image portrayed by a bureaucratic attitude towards policy problem, which has been nurtured by a long-standing juridical tradition and the most narrow-minded administrative studies, as well as by some simplistic top-down interpretations of the implementation process. We do not argue that this kind of complaint is always misplaced, but our previous analysis and the fact that other regions exhibit better performance with the same administrative and juridical architecture cast some doubts on this matter. A management perspective and, not surprisingly, the possibility of adopting a more inclusive participatory style, contrast with the prevailing legalistic rationality displayed by these respondents (note the negative scores obtained on these dimensions).

The opposite frame underlies the answers given by policy-makers operating in Emilia-Romagna. The 'citizenship' factor receives here the highest appreciation, though even in Lombardia the scores are positive. This should mean that institutional decision-makers recognize that the outcomes of the process which they formally control depend equally upon external factors, as on their ability to induce cooperative behaviours from policy-takers. Shortfalls and implementation deficits are mostly caused by lack of communication, incompetences in activating new external resources (and not in consuming existing internal ones), difficulties in confronting typical public goods problems.

In Emilia-Romagna, this attitude is coupled with an equal concern towards what has been named the management frame: efficiency in a world realistically populated by complex problems and private interests. Once again, it is not a surprise that this frame, particularly where it integrates the one defined by a 'collective concern', is not at all worried by possible interferences of an attentive public, not even if it is composed by (potentially hostile) environmental associations. Together with the indefinite attitude shown towards the administrative conception, respondents from Lombardia differentiate themselves from the ones working in Emilia-Romagna for not taking, on average, a clear position on even this last factor.

We have controlled these propositions and their effects in three different ways. First, we have verified, through case studies reconstructions, if the findings of the survey are confirmed by micro, qualitative, context-dependent analyses. In fact, although we can not examine the point thoroughly here, the evolution of the policy area in the three regions and the review of a few 'policy narratives' do not contradict our findings (Giuliani, 1997). Second, we crosschecked them internally; using the same factor analysis technique (applied to a set of statements regarding the preferences and the perceived importance of different policy instruments[13]), we verified that the same geographical differences applied as well to the solution side, and not only to the definition of the problem. Finally, because the extracted factors imply opposite conceptions of the roles assigned to public and private actors, we tried to check their external consistency with the answers given to the questions regarding which type of networks they had activated. It is on this last point that we will now focus our attention.

In Figure 3.2, we presented a set of measures of the general mobilization of actors around the air pollution problem, leaving aside the question of geographical demarcation. Here the problem is that of introducing territorial distinctions. As anticipated, our data do not permit a formal network analysis; none the less, it is possible to delineate the characteristics of the local policy communities with reasonable accuracy, by introducing comparable indicators and thresholds regarding the institutionalization, intensity and complexity of the established web of relationships. Table 3.3 summarizes the available empirical evidence on this matter.

By 'institutionalization' is meant the degree of stabilization, regularization and, possibly, differentiation presented by the relationships studied. For

Table 3.3   Measures of policy networks

Regions	Institutionalization		Intensity	Complexity
	High (> 50% monthly)	Medium (> 20% monthly)	Weekly contacts (at least 3 actors)	No. of contacts (at least 4 issues)
Lombardia	4	6	38.5%	5
Emilia-Romagna	4	6	54.5%	7
Campania	2	4	0%	4

Source: Own survey (Giuliani, 1997)

indicators of this aspect of the policy network, we have observed the number of actors which have been regularly contacted by the respondents of the survey, that is that have established at least monthly contact with them. In cases where at least 50 per cent of the respondents declared regular contacts with a particular actor (for example, the Minister, or environmental associations), we have rated that connection as highly institutionalized.[14] A medium degree of institutionalization has been defined as at least 20 per cent of the respondents mentioning regular contacts. Since the range of actors quoted in the survey included ten different subjects, both indicators have a minimum value of zero (no institutionalization), and a maximum value of ten (all relations are, respectively, highly or medium institutionalized).

From Table 3.3, we can observe that the level of a network's institutionalization is clearly geographically based. Whereas both thresholds are quite steadily surpassed in Lombardia and Emilia-Romagna – each presenting four highly regular connections, and six medium ones – in Campania, institutionalized relationships seem to be restricted to a smaller number of actors. This difference is confirmed if we set up an intuitive 'average score' on all the ten possible associations, by computing the mean of the percentages of regular (at least monthly) relations. This yields a score of 39.6 for Lombardia and 36.6 for Emilia-Romagna, but only 23.3 for Campania, thus confirming a narrower range of institutionalized ties.

With the 'intensity' of association measure, we intend to examine the spread of very intense relations (at least weekly) held at one time by respondents with a suffficient range of other actors (minimum three). Whereas the indicator of institutionalization ranked the 'type' of regular associations, this one evaluates the number of solid webs maintained simultaneously by respondents. Its range goes from a minimum of zero (no respondent has weekly encounters with at least three other policy actors), to a maximum of 100 per cent (they are all tied in very firm networks). Policy-makers operating in Campania confirm the image of isolation already pictured, placing themselves right at the lower limit of the scale. In fact, they seem to respect the typical juridical and administrative division of labour, with institutional actors maintaining regular relations almost exclusively with the level of government immediately above and beneath (for example, provinces with the region and with the municipalities). The situation is clearly different in the other two regions where (as displayed in Table 3.3) the webs of relations seem much more wide reaching. In fact, in those areas, the intensity indicator reaches 38.5 for Lombardia, and 54.5 for Emilia-Romagna.

Finally, let us turn to the concept of 'complexity', with which term we designate those relationships that have been established and are kept up for multiple reasons, that is on diverse issues and problems (but all in the field of air pollution). Since the survey included a specific question on this point, we have been able to ascertain whether each relation had only a small number of aims (as it is the case with the more formal kind of connections), or whether it embraced multiple topics (as with flexible links). To distinguish between these two types of association, we fixed a threshold of (at least) four diverse issues before counting a relation as complex. Once again, our rough index varies from zero (no pair of actors presents this level of complexity), to ten (at least one respondent for each of the ten quoted policy-makers declares multiple-issues relations). In Lombardia, the index positioned itself at the median value of 5, whereas in Emilia-Romagna it scored 7 and in Campania 4, confirming the familiar rank among types of activated networks.[15]

CONCLUSION: SOFT INSTITUTIONS

To summarize the findings reported in the preceding section, the cognitive frames which inform the problem definition and the solution delivery advanced by policy-makers differ significantly among the three regions studied. The survey portrayed with reasonable accuracy the contrast between two opposite policy models: the first, formal, closed, directive and mainly focused upon organizational shortcomings, which appeared to characterize mostly the cognitive maps of the institutional policy-makers employed at different levels of government in Campania; the second, more informal, inclusive, consensual and concerned with policy-takers' reactions, found greater support in Emilia-Romagna. Not surprisingly, besides displaying different policy performance and cognitive models, these two regions are even characterized by opposite types of policy networks. They differ in the number of actors involved, in the degree of institutionalization of the relationships, and in intensity and complexity of the connections in a way that corresponds to the deductive speculations proposed at the end of the fourth section.

Besides looking for the reasons of policy success and failure, we think that one of the merits of the research project outlined above has been that of empirically demonstrating the link between the cognitive models to which policy-makers locally conform, and the characteristics displayed by the network of relationships they tend to activate. The title of this chapter refers to the mix between these two elements using the term 'soft' institutions. The reasons for adopting this terminology should by now be clear.

Networks are certainly not as solidly constructed as more formal institutions such as the parliament or the executive, but they perform similar functions. They stabilize behaviours, establish a friendly environment for mutual recognition through time and space, shape preferences, nurture policy ideas, fix rules of appropriateness, give meaning to the actors' actions. The cognitive frames which we have discussed and analysed in the preceding sections evolve in, and emerge from, policy networks. They are embedded in them. Decision-makers only think within their limits; they think the thinkable (Faure, Pollet and Warin, 1995). At the same time, what keeps policy actors continuously floating around in this whirlpool of relations is the fact that they partake in a common concern, they share ideas. It is precisely this cognitive glue that defines the boundaries of networks, who is in and who is out, the limits for the exchange of information. Loose ties, and not formal rules, frame the basic nature of soft institutions.

As far as the air pollution problem is concerned (and briefly turning to some normative argument), the empirical evidence presented here is strongly in favour of a more elastic approach to policy regulation. Beyond a minimal organizational equipment, effectiveness and performance do not seem to be correlated to the amount of resources (of different types) available to local administrations. As has been pointed out by some of the earliest implementation studies, the attitude of the agency in charge is a key element of the whole implementation process, and an important factor for its success. Once again, the concept of a 'soft' institution – operationalized through the analysis of the definition of the policy problem given by policy-makers themselves, and through the web of relations they tend to activate – has proved its usefulness in this matter.

To put it all into a nutshell: open networks, favouring informal but intense and complex relations within a wider range of actors, and fostering a problem definition which contrasted a rigid administrative perspective while promoting the potential contribution of cooperative citizens and policy-takers, appear to assure the highest guarantees for an effective implementation of anti-pollution programmes. Public participation may not be of value in itself, but a consensus-building attitude on the part of institutional policy-makers may represent a good starting point for positive-sum processes. This is certainly not a universal solution, nor the policy answer to every kind of problem. The collective character of the air pollution issue itself, which is extremely sensitive to the reactions of policy-takers – in using public transport, reducing heating when possible, paying attention to one's car emissions, and so on – is probably at the root of the success of the described solution. Building soft institutions of the kind dealt with in this chapter; though not necessarily expensive or politically puzzling, remains a hard task to accomplish. But a praiseworthy one.

NOTES

1. For a preliminary ordering of the policy network category see Jordan and Schubert (1992). For an introduction on the importance of cognitive variables in the policy process, see Goldstein and Keohane (1993), Majone (1989) and Stone (1988).
2. We had the opportunity to calculate the trend in the approval of environmental laws in the fifteen Italian ordinary regions, from their establishment, in the early 1970s, to 1994. The absolute number and the relative proportion of environmental laws constantly increased through the years that preceded the electoral appointment, usually peaking in the electoral year (occasionally the year before), and sharply decreasing the year after (Giuliani, 1997). The typical public-good content of environmental regulations may be the key to understanding this impressively recurrent politico-environmental cycle.
3. We have verified the absolute and relative amount of environmental laws adopted in each region in the period 1970–94, whereas the implementation capacity has been determined starting from a cumulative index computed from fourteen indicators of conformance to national prescriptions (Legambiente, 1995), and from five more variables addressing the issues of environmental financing and spending, and the magnitude of the phenomenon of ecological infractions. See Giuliani (1997) for a more accurate description of both indexes.
4. For the economical resources we calculated the amount of funds per head provided for air pollution by the environmental plan 1994–96: these data included the financing of regional plans for restoring the air quality, money for developing systems of pollution monitoring, and incentives for the introduction of traffic measures and the adoption of less polluting transport systems. Regarding staff resources, we examined the number of personnel employed at the beginning of the 1990s by the PMPs (Presidi Multizonali di Prevenzione), the main technical agencies of the Health Service which locally supported the management of anti-pollution policies. Under normative resources we included a qualitative evaluation of the introduction, institutionalization and consistency of the acts locally adopted, as well as their conformity with the planning obligations fixed at the national level. Finally, regarding what we have called information resources, besides the existence of information channels to and from the citizenship, we estimated the number of existing air pollution monitoring stations per head, and per urban area.
5. Emissions have been evaluated through the 1980s for five types of pollutants (SOx, NOx, CO, particulates and volatile organic compounds). The availability of NO_2 measures of air quality has been considered an index for itself, besides including its actual trend in the early 1990s. Lastly, the percentage of the population using public transport has been examined as an indicator of effective sustainable policies in the field of air pollution.
6. This finding obviously cannot be interpreted in any prescriptive sense (such as 'to improve outcomes, we should reduce the resources'), but only as a hint in the direction of different explanatory variables. We can add that a parallel analysis of the solid waste policy gave a higher correlation coefficient ($R^2 = 0.49$). The minor role played by organizational resources in the air pollution case may be attributed to the technical complexity of the problem, and to the major influence of contextual factors. In the next section we will suggest a research strategy which should be able to grasp new elements in this direction.

7. By 'institutional policy-makers' we mean those officials responsible for air pollution controls inside regional, provincial and municipal administrations. The adjective 'institutional' is used simply to distinguish these actors from environmentalists, experts, technicians and other kinds of potential and actual policy-makers. The most sophisticated statistical method used was factor analysis which, beyond being intuitively comprehensible, does not require a large number of respondents.
8. The air pollution problem is not confined only to the main towns, but extends to most of the provinces and to the other big cities. For a review of the problem, see Giuliani (1997) and Lanzalaco et al. (1996).
9. This happened because the number of policy-makers itself was smaller in Campania, which has only five provinces, than in Emilia-Romagna (nine provinces) and Lombardia (eleven provinces).
10. Policy-makers rated as 'most problematic', 'quite problematic' or 'unproblematic' the eighteen statements reported in Table 3.2, regarding the puzzling aspects of the air pollution policy. The three factors extracted – which were calculated on all our respondents, and not *ad hoc*, only on the answers obtained from the three regions investigated in depth – account for almost 50 per cent of the original variance.
11. The names assigned are somewhat arbitrary, because it is the underlying composition of factors that matters. The relevance of the extracted dimensions, and their non-contingent character, has been partly confirmed by finding similar underlying cognitive maps in the area of solid-waste policy (Giuliani, 1997). This demonstrates a certain degree of geographical stability of the interpretative frames adopted by policy-makers and, at the same time, the possibility of a generalization effort.
12. In Sartori's terms (1987, p. 214), collectivized decisions are choices which 'whoever does the deciding, decides for all'. The defining criterion is the scope, not the range of actors envolved. Collective decisions, on the other hand, 'are understood to mean decisions taken by "the many" '.
13. These concerned twelve policy choices, which the factors analysis grouped into three intervening strategies mainly addressed to: 1) transportation; 2) traffic; 3) control of emissions.
14. Regular contacts can be established for personal or entrepreneurial reasons; in which case they cease naturally in the course of staff turnover – which is a sign of poor institutionalization. The reason for fixing a 50 per cent threshold was exactly to avoid including that kind of personal (though regular) contact.
15. Even by computing the exact average of issues tackled for each relationship, the policy-makers of the central region scored best, though the southern ones improved their relative position.

4. Improving air quality in Italian cities: the outcome of an emergency policy style*
Carlo Desideri and Rudy Lewanski

Dependence on private vehicles for the transportation of passengers and goods in Italy has reached levels that are among the highest in Europe. The reason mainly lies in the lack of investments for the development of modern public transportation systems, (such as efficient and reliable railways), both within urban areas and between them and more generally in the lack of any transport policies, 'let alone any coherent policies, that causes a chaotic urban transport situation'. (Pucher and Lefêvre 1996, pp. 4 and 101). Thus individuals have not had many alternatives to the road in the choice of means of transportation. Inside urban areas, the lack of adequate infrastructure such as roads and parking facilities, coupled with high levels of traffic, has given rise to enormous negative externalities (congestion, pollution, accidents and the like). This chapter's aim is to describe the main features of the policies adopted in this area both by the central government and by local authorities. With the exception of measures adopted by a few cities in the early 1980s, public policies have been mainly characterized by a reactive style, however, they have at least been effective in producing social learning processes and asserting the legitimacy of the air pollution topic to be included in traffic policy debate.

MOBILITY AND ITS EFFECTS IN LARGE URBAN AREAS

In Italy, more than in other countries, a steep increase in mobility has meant a strong shift towards road transport for both freight and passenger traffic (see Tables 4.1 and 4.2). Between 1970 and 1991 road traffic increased by 132 per cent, whereas GDP rose by 83 per cent (OECD, 1994, p. 125). At present 38.2 million vehicles (29.6 million cars, 2.6 million trucks, 5.8 million motorcycles) are circulating on Italian roads. Italy ranks second in Europe after Luxembourg in terms of the number of vehicles per inhabitant. Together with Germany, it has passed the mark of one vehicle for every two persons (LegAmbiente, 1996,

Table 4.1 Prevailing mode of passenger transport in Italy, 1970–93 (percentage of total)

Transport mode	1970	1980	1985	1990	1993
Car/motorcycle	76.1	76.1	77.4	79.7	82.0
Railways	11.4	9.4	7.7	6.6	6.1
Air	0.4	0.6	0.8	0.8	0.8
Other public transport*	9.2	9.1	7.5	4.8	3.9
Sea	0.3	0.3	0.3	0.3	0.2

* Excludes tramways

Source: ISTAT (1993), p. 227

p. 104); there are 1.5 inhabitants per car in Florence, 1.6 in Milan and Rome, and 1.7 in Bologna. If one considers the situation in relation to available space, Italy appears to be more congested than other countries with 94 vehicles per kilometre of existing roads – almost double the European average of 49 (Ministero dell'Ambiente 1992, p. 337). The situation in large urban areas is even worse: Milan has 5362 vehicles per km^2, Turin 4839, Naples 4380, Florence 2466, Bologna 1772, and Rome, thanks to the size of its municipal territory- only 1379. These figures do not include the traffic regularly coming from surrounding municipalities, but only that of residents; the daily influx of people coming in from other municipalities varies from 27 per cent in the case of Milan to 13.8 per cent in that of Turin (*La Repubblica*, 8 October 1996, p. 30; LegAmbiente 1996, p. 231). In addition, the increase in the number of vehicles has been sharper in Italian cities (51.2 per cent in Rome, 63.5 per cent in Naples between 1970 and 1990) as compared to other European countries (for

Table 4.2 Prevailing mode of freight transport in Italy, 1970–93 (percentage of total)

Transport mode	1970	1980	1985	1990	1993
Lorry	51.8	65.8	71.2	71.9	72.7
Railway	16.7	10.5	9.3	8.8	8.4
Water	23.4	17.2	14.8	14.4	13.8
Pipeline	8.0	6.4	4.4	4.6	4.9

Source: ISTAT (1993), p. 227

example, 6.7 per cent in London, 24.1 per cent in Barcelona; Lizzi et al., 1996, p. 5). Vehicle congestion is connected to the high population density in large cities (9102 inhabitants per km in Naples, more than 7400 in Turin and Milan) which is high as compared to European averages. Besides vehicle ownership, the actual use of private vehicles has increased – one effect of the continuing trend for inhabitants to move away from city centres towards suburbs driven by high housing costs.

In urban areas public transport has steadily lost ground *vis-à-vis* private vehicles. Between 1981 and 1991 the use of public transport (road and rail) declined by approximately 25 per cent (Ministero dei Trasporti, 1995a, p. 43). While reliance on cars in urban areas almost doubled in only 8 years, increasing from 117.4 billion km in 1985 to 225.8 in 1993. During peak hours only 45 per cent of personal mobility takes place by public transportation in Milan, and 35 per cent in Naples, as against 55 per cent in Paris, 70 per cent in Munich and Vienna, and 85 per cent in London (UITP-IVECO, 1994). According to a survey carried out in 1991, only 16.3 per cent of students and workers use public transport, whereas 41.6 per cent use a private motor vehicle (plus 9.4 per cent a motorcycle) and the remaining 32.5 per cent walk (Ministero dei Trasporti, 1995a, p. 43). The average commercial speed of surface public transport is around 3–4 km per hour, and there are only 104 kilometres of underground transport system (mainly in Milan and Rome) (Ministero dei Trasporti 1995b, p. 64).

This extensive reliance on private vehicles has substantially contributed to the deterioration of air quality in cities. Until the 1970s, the main cause of air pollution – a serious health issue, particularly in northern cities because of local meteorological factors – was from stationary sources (heating and industries). Environmental policies[1], especially energy-related policies (for example the change to cleaner fuels[2]), substantially reduced emissions from such sources. Global emissions of traditional pollutants such as SO_2 have declined over the years and air quality standards have considerably improved for such substances (OECD, 1994, p. 163). On the other hand, vehicle pollutants have steadily increased over the years; traffic now accounts for 49 per cent of NOx emissions, 63 per cent of CO, 38 per cent of Volatile Organic Compounds (VOCs) 62 per cent of lead and 20 per cent of CO_2. In relation to total vehicle emissions, urban traffic alone accounts for 77 per cent of CO emissions, 39 per cent of CO_2, 27 per cent of NOx, 76 per cent of VOCs (plus another 5 per cent from gas stations) and 29 per cent of total suspended particles (Ministero dell'Ambiente, 1996, p. 32).

The concentrations of air-borne lead are certain to slowly diminish; unleaded petrol at present represents about 34 per cent of total sales. This figure, however, is still considerably lower than the EU average of 64 per cent (LegAmbiente, 1996, p. 120). In addition, use of unleaded petrol by vehicles not equipped with a catalyser (still 80 per cent of total) because of the price differential as compared to leaded petrol is producing increasing levels of other pollutants such as benzene.[3] According to data provided by the Green Trainí promoted by the NGO LegAmbiente, the level of benzene in urban areas is 15 times higher than the levels specified by the decree of November 1994.

Obviously, the negative effects of vehicle circulation are not confined to air quality. In 1994 more than 170 000 accidents occurred, causing 6578 deaths (9000 including deaths occurring more than one week after the accident) and 239 000 injured persons; 73 per cent of accidents take place in urban areas (causing 42 per cent of total accident deaths and 70 per cent of injuries). The accident rate is considerably higher than the European average (5.5 accidents per 10 km of roads as against 3.3 average; 2.3 victims per 100 km of roads against 1.4 average). The 72 per cent of population living in urban areas is exposed to noise levels, mainly due to vehicular traffic (Ponti and Vittadini, 1990, p. 197), that exceed maximum thresholds specified by European and national regulations (65 dBA Leq by day and 55 by night). Also the annual cost of traffic congestion is considerable: in Rome alone it was calculated as 1600 billion lire in 1987 (Ponti and Vittadini, 1990, p. 201).

NATIONAL POLICIES

Local traffic policies in Italy cannot be understood without an overview of national policies that deeply affect the former.

In the field of vehicle emissions, Italian governments in the past have limited themselves to enacting standards established by UNECE in the 1960s and by the EC since the 1970s, (although it has played an active role in defining the latter during the protracted decision-making process that took place in the 1980s (Lewanski, 1997, pp. 75–6 and 225–33).

Air quality standards for SO_2, NOx, lead, fluoride, CO, O_3 and particulate matter were introduced for the first time by a decree of 28 March 1983, according to which all regions were supposed to draw up clean-up plans by 1993. Subsequently, Decree no. 203 of 24 May 1988 enacted various EC Directives containing standards for such pollutants as SO_2, NOx, particulate matter and lead.

In November 1991 the Minister of the Environment (Ruffolo), in conjunction with his colleague responsible for Urban Areas, and on the basis of powers invested in him by Law no. 59/87, to intervene in urgent and exceptional cases to protect health and the environment, issued eleven ordinances containing emergency measures aimed at dealing with air and noise pollution in the major urban areas of the country for a limited period of time during the winter of 1991/2). It should be noted, however, that these powers were not being used in emergency conditions, but rather in a chronic situation that had existed for years and was not at all exceptional. The real emergency was represented by the lack of adequate policies and provisions and by the inertia of local administrations in this field (Desideri, 1993, p. 28).

The emergency measures specified alert and alarm thresholds for a number of urban air pollutants. On the basis of such thresholds, mayors were to impose restrictions on civilian heating, industrial activities and traffic circulation in order to limit emissions. Many cities at that time did not have equipment capable of measuring pollution, thus the ordinances also contained emergency provisions for setting up monitoring stations. Financial resources had already been allocated by the Ministry for this purpose, but these had not yet been used. A decree issued in May 1991 specified the number of units to be set up in each city according to the population (six for up to 500 000 inhabitants, eight for up to 1.5 million and twelve beyond that threshold). The Ministry also laid down the characteristics of fuels to be used in the same eleven areas in terms of content of sulphur, benzene and PAHs. This measure drew complaints from the oil companies that they had not been consulted and, as a consequence, the Minister agreed to a compromise with the companies on the basis of which the benzene limit was established at 2.5 per cent as a monthly average (rather than as an absolute value).

The ordinances caused considerable conflict with local governments. The association of municipalities (ANCI) accused the government of centralist and authoritarian behaviour, and complained that they had not been consulted before the Ordinances were issued. Such protests persuaded the Minister to modify his provisions, allowing for more leeway in the choice of the measures to be adopted by local authorities. Though the governmental provisions were seen by local governments as unlawful interference in their affairs, it should be noted that, as these measures had been imposed by the Ministry, by attributing the responsibility to Rome they could, at least partially, escape paying part of the political price of the unpopular measures adopted. Imposing losses, as noted by Pierson and Weaver (1993, p. 110), is politically difficult 'because doing so incurs costs that are concentrated, immediate and highly visible, while the benefits are ... diffuse and long-term'; politicians are motivated 'by the desire to avoid blame for unpopular actions since voters are more sensitive to losses than they are to gains' (Weaver, 1986, p. 371).

In 1992, the new Minister, Ripa di Meana (previously EU Environment Commissioner), after consultations with other ministries and local authorities, issued a Decree on 12 November that tried to provide a permanent regulation on this matter (though the legal basis of the decree was questionable; Desideri, 1993, p. 18). The Decree applied directly to fifteen cities (the previous eleven, plus Modena, Reggio Ernilici, Lucca and Pescara, that had asked to be included on a voluntary basis), and the regions could extend its application to other highly polluted areas provided some type of measurement device existed within them.

The Decree set out a large number of possible measures, from which each city could choose which to adopt. Local authorities were to be assisted by a technical body, formed by local experts of their own choice, in the analysis of the pollution data and the choice of measures to be taken. The Ministry also reached a voluntary agreement with the oil companies in October 1992, fixing the maximum benzene content of diesel oil at 3 per cent and that of sulphur at 0.2 per cent.

The Ripa di Meana Decree was abolished in March 1994 by a decision of the Constitutional Court upon the opposition by the Lombardy Region. A new provision issued first on 15 April of the same year by Minister Spini, and subsequently modified by another Decree issued by Minister Matteoli on 25 November, is less stringent as far as the limits of urban air pollutants are concerned and leaves considerable leeway to local authorities (Zolea, 1996, p. 90). However, the same provision requires the local authorities of 23 cities with more than 150 000 inhabitants to monitor benzene, PAHs and PM10, and lays down stringent air quality limits for such pollutants, that must be observed from January 1996, with even lower limits from January 1999.

With regard to the content of benzene in fuels, a law by decree issued by the Government in September 1995 proposed a limit of 1.4 per cent as of July 1997 and 1 per cent as of July 1999 (the EU target for the year 2000); however, this law has never been ratified by Parliament. In 1995 the average benzene content of fuel was 1.6 per cent and the PAHs content 33.8 per cent (Unione Petrolifera, 1996). One company, AGIP, recently launched a campaign advertising its petrol as containing only 1 per cent of benzene (*Il Sole 24 Ore*, 7 June 1996). Unleaded gasoline continues to be less expensive than leaded (about 100 lire less per litre), though such preferential detaxation, initially aimed at promoting the sales of cars with catalysers, makes little sense now all new vehicles are equipped with such devices as a result of EU legislation (Buglione, 1996, pp. 150–51).

The interest of the central authorities in pollutants such as benzene and PAHs is explained by the increasing quantities being emitted as the consumption of unleaded, so-called 'green' petrol by non-catalysed cars continues to rise.[4] The national Toxicological Commission pointed out in 1990 the carcinogenic danger – recognized since the mid-1980s – implicit in the growing diffusion of such substances. (The Commission's advice to forbid the sale of unleaded petrol to non-catalysed vehicles went unheeded.) A Turin judge, well known for his activism in the field of environmental and workers protection, issued an indictment against the oil companies on the grounds that a high benzene content represents a potential menace to the health of personnel working at petrol pumps. The very slow turnover of the vehicle fleet means that any positive effects of catalisation will begin to be felt only in the next century (provided that vehicles are maintained in good condition).[5]

The new Road Code passed in 1992, which superceded that of 1959, provides for periodical emission controls as a result of the EU Directive 55/92. The new legislation provides for general checkups, initially four years after manufacture and every two years thereafter (previously after ten years for new vehicles, and then after every five years), and allows local authorities to establish further emission controls. Such activity will have to be carried out by private authorized garages since the competent provincial Motorizzazione Civile could not possibly handle the huge amount of work involved.[6] This might also be seen as a way for the central government to offer indirectly an incentive (without having to bear financial expenditures) to vehicle sales that, due to the poor economic situation in Italy, have been depressed over the last two years. It should be noted that 39 per cent of cars in circulation are over nine years old (Ministero dell'Ambiente, 1996, p. 26) and this represents a serious contribution to pollution since emissions increase sharply after six years of age. Furthermore, the Prodi Government launched a successful programme encouraging car owners to trade in vehicles 10 years old or more (to be scrapped) for new ones; this programme, based on a financial incentive that has to be matched by vehicle manufacturers, was officially aimed at helping the car industry as well as improving environmental conditions (on the assumption that new vehicles are less polluting).[7]

The new Code also allows local authorities to adopt a number of measures which, in fact, they were to some extent already carrying out, such as establishing areas where traffic is limited to specific categories of vehicles in order to protect environmental conditions and cultural heritage (monuments and the like), pedestrian areas, bus lanes, streets reserved for public transport, and road pricing. It also finally permits municipalities to request payment for parking even when vehicles are not supervised (thus offering municipalities a source of local income). According to the law, however, any profits must be invested in traffic management measures, such as parking. All municipalities with more than 30 000 inhabitants are supposed to pass two-year urban traffic plans (PUT); limitation of air (and noise) pollution is explicitly mentioned among the objectives of such plans.

Finally, it should be noted that, although Italy's petrol taxes are among the highest in Europe (OECD, 1994, p. 133), they have been used merely as a source of revenue for central and, more recently and to a limited extent, regional government, rather than as a tool for limiting vehicular traffic[8] (Buglione, 1996, pp. 132 ff.). Registration and ownership taxes increase progressively according to engine size (but not to actual emissions). Old diesel cars are severely taxed, making them convenient only for high mileages.

LOCAL POLICIES FOR AIR POLLUTION REDUCTION

Command-and-control Measures

Traffic restrictions in sections of cities

Such measures have been introduced with two purposes that are conceptually distinct, although in practice difficult to separate.

The first goal was to reduce congestion, especially where density is highest and urban features least suitable for vehicle circulation (typically old historic town centres). Naples, for example, introduced limitations to circulation on the basis of even and odd licence plates on alternate days in an attempt to contain congestion; Bologna closed most of its centre (350 hectares of a total of 440) for the entire day (7 am–8 pm) in 1985 (however, some 50 000 permits were issued to residents, owners of parking places and economic operators including tradesmen on call, couriers, doctors and salesmen); and shortly after Bologna, Florence also created a large restricted traffic zone.

In order to enforce its decision to close its historic centre, Bologna also introduced, for the first time in Italy, an electronic control system (Sirio) capable of automatically reading the licence plates of vehicles entering the central zone and issuing fines to the unauthorized ones. The advantage of this system is that it would free a considerable number of policemen who would otherwise have to be stationed at the entrance points of the historical centre. Even though the Sirio system was financed by the Environment Ministry, only after it was installed did it become evident that the new system also required the approval of the Ministry of Public Works (in charge of traffic issues). The approval was delayed for two years, probably owing to bureaucratic inertia, but also because of political 'sabotage' on the part of the rightist Berlusconi government in opposition to the leftist coalition governing Bologna. Since the new and politically more sympathetic Prodi government took power, Sirio is being rescued from the authorization quicksands and a number of other cities (such as Rome) are now considering the introduction of a similar system.

The second goal of traffic reduction measures was to contain pollution levels. Hitherto, such measures have been adopted reactively, for example when alert and alarm thresholds are exceeded (on the basis of the Decrees issued by the Environment Ministry which we have already discussed). It should, however, be noted that there exists considerable leeway for the town authorities in deciding how to actually apply traffic limitations; for example, under such relevant aspects as timings, categories exempted, or effectiveness of controls. In 1993 the Mayor of Milan in practice nullified the Ripa di Meana Decree by allowing anybody who claimed the need to do so to use their own vehicle: the consequence was 'business as usual'.

Banning vehicle circulation after high levels of pollution are reached has met with high unpopularity because of the immediate disruption of everyday mobility patterns. Furthermore, the actual effectiveness of such measures has proved to be quite low so far as air quality improvement is concerned (CSST, 1994). Such factors are persuading city authorities to approach traffic restrictions on a pre-planned basis using historical trends for the most polluted months of the year or projections of pollution and climatological data that allow them to predict with some accuracy when high levels of pollution will occur (Turin, Florence and Bologna have already adopted such an approach or are in the process of doing so).

Many cities have created pedestrian-only areas: in Bologna some 30 000 m^2 in the centre have been completely set aside for pedestrians since the 1970s; Rome is pursuing a strategy of pedestrian 'archipelagos'; Milan closed one of its most fashionable town-centre shopping streets 1995. It should be noted, however, that the purpose in such cases is to favour leisurely shopping rather than limiting traffic *per se*.

Emission controls

Modena was the first city to introduce compulsory periodical (usually annual) controls of vehicle emissions, followed since 1994 by Milan, Turin and Bologna. Some cities (Rome in 1993) initially attempted a voluntary approach but when that failed the administrations eventually turned to compulsory controls (which the national legislation discussed above allowed them to do).

Planning activities

As mentioned earlier, planning activities in respect to air pollution generated by traffic take place both at the regional level, through clean-up plans provided for by Decree 203/83 (prescribing actions to be taken in order to improve air quality), and at the local level through Urban Traffic Plans (PUT) provided for by the 1992 Road Code. However, only about one-quarter of the regions had approved a clean-up plan by the 1993 deadline (LegAmbiente, 1996, p. 202), and only 18 per cent out of a total of 587 municipalities actually introduced the PUT plans by the end of 1996 .

Economic Measures

Until recently, road space was allocated free of charge for vehicle parking. The first attempts of municipalities to introduce parking meters clashed with jurisprudencial interpretation of legislation according to which payment could be requested only if vehicles were guarded. The new Traffic Code, as mentioned previously, has finally authorized payment for unsupervised parking. Thus, in order to be able to collect financial resources too, many cities have started to introduce parking meters or 'vouchers', and other systems that compel vehicle users to pay for roadside parking. Turin represents a notable example, having introduced a quite effective payment scheme in the city centre (the 3 billion lire profit that was collected during the first year was used to build a new parking facility); many other cities have already, or are considering, payment schemes for parking in more or less extended areas (usually with tariffs increasing as one gets nearer to the centre). Rome, Bologna and Perugia will soon adopt similar schemes. Udine (the first city to start in the early 1990s), Siena, Vicenza, Mestre and Padova require even residents to pay a fixed monthly sum in order to have the right to park on the street.

Road pricing schemes are also starting to attract the interest of municipal authorities. Como, a small city located in Northern Lombardy, has obtained financial assistance from the EU towards setting up such a system, and other cities including Rome and Florence are considering the possibility of introducing one. It is also worth mentioning that regions can decide variations on the price of petrol (though in fact they have never done so until now); such variations, however, are so small (30 lire/litre) that they would not be able to significantly influence the use of vehicles (Buglione, 1996, p. 159).

Infrastructural Measures

Parking facilities

A national law passed in 1989 allocated substantial resources (2000 billion lire) for the construction of parking facilities, but until now it has hardly been a success. Only 25 per cent of the 140 000 places to be built in large cities have actually been completed so far (*L'Unita*, 3 September 1996). Furthermore, the facilities that have been built – at the outskirts of towns, have often remained empty because of the possibility of free on-street parking and the improbability of being fined (COREP, 1994, p. 35).

Public transport

The availability of adequate public transport systems is obviously an essential condition of avoiding traffic congestion and high levels of pollution. This necessity, however, has been one of the major weaknesses of local traffic policies. Though there has been a general trend towards the increase of car use in most European cities since 1970 (Pucher and Lefêvre, 1996, p. 16), the poor performance of surface public transport – one indicator of which is an average age for buses of 12 years, among the oldest in Europe – has further induced many Italians to prefer private vehicles. After all, if one has to pass long hours in traffic jams, it is certainly more comfortable to do so in your own car than in an overcrowded bus! The result is that the percentage of people using public transport has continued to decline over the years.

In an attempt to improve the situation, or at least to counter the constant loss of terrain *vis-à-vis* private traffic, some cities (Bologna, Turin, Naples and Rome) have introduced measures aimed at increasing average commercial speed (such as bus, tram and taxi lanes) or, more recently, electronic devices capable of assuring priority at junctions (by changing the light to green as the public vehicle approaches for example, in Modena). Rome, Milan and Bologna have created integrated tariff systems that allow people to use transportation services managed by different authorities. Other measures, such as creating parking facilities served by public transportation at the periphery of cities, have met with very limited success, mainly due to the expectation of drivers to find free parking nearer to the centre. Bologna tried a more radical approach between 1972 and 1976, when free travel on public transport was offered even during rush hours to encourage its use; this, however, was also conceived of as a welfare policy, and in any case was abandoned due to financial constraints.

Trams and trolley buses were gradually withdrawn from many cities (for example, Bologna, Florence) during the 1950s and 1960s and replaced by more 'modern' and flexible buses, without concern for environmental aspects. Only a few cities (chiefly Milan and Turin) were wise – or perhaps only inactive – enough to retain some of their tram lines. The 665 kilometres of rails present in 1970 were reduced to 405 by 1993 (Ministero dei Trasporti, 1995b, p. 63). In order to improve public transport, a number of cities have aimed at building underground systems. Milan currently has three lines, Rome two and Naples one; the construction of lines in other cities has recently been financed by the central government. Many cities (Florence, Bologna, Rome) and regions (Emilia Romagna) have used the opportunity of granting authorizations for the construction of the high speed train (TAV) to bargain for investments in the field of local transportation by the railway companies and the use of existing tracks for metropolitan transportation in exchange for such authorizations.

In some cities the idea of building underground lines has had to be reconsidered in favour of less expensive surface rail systems. Bologna represents an interesting example of 'social learning' in this respect. During the 1980s the municipality developed a project for an underground system, however, the pressure of the local Green party, together with the lack of financial resources, eventually persuaded the administration to change over to a tramway. A single line costing 400 billion lire, to be covered by the state and the municipality, was recently financed.

Air Quality Monitoring

Some municipalities have dedicated considerable resources to pollution measurement. Before the policies enacted by central government in the early 1990s mentioned above, only a few large (Milan, Turin, Bologna) cities had installed monitoring networks during the 1970s. These were often inadequate because they included a small number of units and measured the more traditional pollutants produced mainly by stationary sources (typically SO^2 and particulate). The measurement of NOx and CO in these cities began later, in the following decade. The intervention of the central government forced cities without measurement stations to install them, and those that already had them to increase their number and the pollutants measured (benzene, ozone). The Ministry of Environment also provided the necessary financial resources. Bologna, for example, recently extended its network in order to cover the metropolitan area and the types of pollutants analysed (benzene, toluene, xilenes and PM10, that the local health officer judges especially dangerous for human health). The case of Modena is indicative of a particularly active administration: though not a highly polluted city, by 1990 it had already set up a network of eleven stations, and subsequently extended it to measure non-traditional pollutants such as VOCs, PAHs, heavy metals and acidic compounds (Lizzi et al., 1996).

At present there are 452 stations (ISTAT, 1996, p. 52), concentrated mainly in Northern and Central Italy (Ministero dell'Ambiente, 1994, p. 16). It should be noted that these figures also include monitoring around industrial and energy production activities in extra-urban areas. Furthermore, many networks present serious operational problems and the issue of data quality assurance still remains to be solved (Zolea, 1996, p. 94; ISTAT, 1996, p. 66). At present, 79 cities and towns carry out some form of air quality monitoring, although very seldom do the measurement stations meet the criteria required by law (such as the minimum number of six monitoring stations). Only 59 of these also measure the levels of ozone in the air and that of benzene (Fiorillo, 1996, p. 36).

Behavioural and Organizational Policy Approaches

Only very recently are cities beginning to attempt to reduce peak-hour traffic congestion and pollution by modifying school and working hours. The first city to experiment with such a solution was Naples; other cities are considering similar measures. Rome set up an *ad hoc* 'Citizen timings' office responsible for analysing and proposing changes in the opening hours of services and shops.

Bologna launched a 'Decalogue' suggesting behaviour motorists should adopt in order to reduce pollution. Such policies have little impact in general, and in Italy – due to the prevailing (un)civic culture – even less so. There were attempts in several cities to promote car pooling; these, however, have met with very meagre success.

Modena reorganized the timings of refuse collection trucks to make traffic more fluid. However the lack of financial resources and resistance from trade unions represent serious obstacles to such efforts. Cities have also met considerable resistance to attempts to regulate the working hours of trucks that impede the traffic flow delivering goods to shops.

POLICY NETWORKS: CONFLICT AND CONSENSUS.

The Policy Arena

The urban traffic policy arena does not seem to feature, differently from other sectoral policies, clearcut, tight and stable coalitions. Perhaps an appropriate metaphor could be that of a policy 'web' on which a number of 'spiders' move around hunting for the stakes of the game (the insects trapped in the web).

Coalitions and levels of mobilization change according to the stakes but also, to some degree, on the basis of other factors such as changing belief systems.

At the centre of the web is the focal actor, the municipality; its importance is due to the decision-making powers entrusted to it. The municipality can hardly be considered a monolithic actor: whereas the Traffic *assessorato*, sometimes supported by the public transportation company (*azienda municipalizzata*), will be in favour of restrictive measures to private vehicles, other *assessorati* will defend the interests of specific categories, such the shopkeepers, the disabled and so on. The mayor typically acts in favour of compromise both among various departments and between these and external interest groups. Briefly turning our attention to the latter, on one side of the web actors in favour of limitations to private traffic are located. Though dissatisfaction with the poor environmental conditions caused by traffic is widespread (as may be gathered from letters published in local newspapers), actual mobilization is more rare. Information strategies by public authorities are insufficient to support such mobilization (Desideri, 1996, p. 65). There certainly have been episodes of protests against traffic and pollution (petitions, gatherings and demonstrations), but these have been short lived and their influence on policy-making appears limited. Ecological groups – WWF, LegAmbiente, Greenpeace, other local associations and *ad hoc* committees – have been more lasting over time, but their voice, though amplified by the media, has been easily submerged by that of affected economic interest groups in the decision-making process. In some cases, as in that of benzene (Zolea 1996, p. 107), scientific actors (health authorities, academics) have tried to point out the dangers arising from traffic pollution. The creation of local technical bodies in some cities, following the previously described Decree issued by Ripa di Meana, has also brought scientific actors into the policy process.

The media – especially the press – have also played an important role in pointing out inadequacies of local policies (Desideri, 1993, pp. 33–7) and in creating a demand for more effective measures and more stringent standards (for example in respect to the levels of benzene).

Certain organized categories, such as bus drivers and municipal traffic police, prompted by their unhealthy working conditions, have occasionally supported traffic restrictions, but their mobilization has quickly faded as well. Other categories, such as taxi drivers who own their vehicles, have been more concerned with safeguarding short-term corporatist interests rather than realizing the long-term economic advantages potentially deriving from innovative policies.

On the other side of the web one finds economic interests –especially shopkeepers – that strongly oppose restrictions to private vehicles for fear of losing potential customers. The only exception concerns the creation of pedestrian streets or areas where the most fashionable shops are located, since shopowners of this type by now well understand the advantages of such measures. Car owners are generally opposed to circulation and parking restrictions, speed restrictions, or measures that take away space from them (for examples, bus lanes). The local Automobile Club (ACI) has often made representations on behalf of this category.

But the web sways and the actors refuse to stay put: attitudes change according to particular circumstances and how individuals are affected by general policies and specific measures. Those in favour of unrestricted vehicle use take up arms in accordance with the NIMBY – not in *my* back yard – syndrome, when a parking facility is to be built near 'their' residence or traffic reorganization measures bring heavy traffic to 'their' streets.

Policies aimed at regulating private traffic in Italy appear to be doomed to encounter substantial resistance and raise acute conflicts. Several episodes reveal the bitterness of the conflicts that traffic control measures are capable of igniting. The traffic police office in Florence was destroyed by arson. In Trieste things went further: shop-owners protesting against the anti-pollution measures resorted to physical violence in an attempt to occupy the municipality and threatened several representatives of the local government; last year the political head (*Assessore*) of the city's Traffic Department who supported restrictive measures in order to limit pollution levels, was actually assassinated. The killer was judged to be insane, but just for that reason he was particularly sensitive to the tension 'in the air'.

Opponents also resort to more peaceful, albeit often rather effective, means that can range from political pressures – as in the case of shop-keeper strikes in Rome and Bologna – to court actions. The administrative tribunals (TARs) have often lent attentive ears to special interests on the basis of very formalistic lines of juridical reasoning. The Emilia Romagna TAR, for example, has been especially active in blocking such decisive measures taken by the Bologna Administration, the Mayor's decision to replace the chief of municipal police because of his poor record in controlling traffic, the decision in 1993 to carry out compulsory checks of vehicle emissions (Bologna was only able to reintroduce such measure in 1996 after national legislation was passed); and the functioning of the Sirio system described above.

Local authorities are obviously keen to avoid conflict as far as possible, partly because they realize that effectiveness depends on the perceived legitimacy of the adopted policies,[10] but more especially because, as mentioned above, they fear paying the political price inherent in unpopular measures. Solutions based on the construction of infrastructures that have distributional effects are seen as less politically risky than regulatory measures. To complicate things further, traffic represents a typical example of an issue that to a large extent cuts across traditional political divisions and party alignments.

Thus, it is not surprising that authorities have given their policies a consensual slant. Most cities have attempted to ensure agreement on measures adopted and created opportunities ('tables', forums in Rome and Bologna, and committees) for various interest groups to provide their input before final decisions are taken. An interesting example of a cooperative approach with affected parties is represented by the 'Mobility Pact' promoted by Modena in 1994, which included a number of projects (promotion of public transportation, desynchronization of working hours, school and shop hours, car pooling, and so on) agreed upon by economic, social and institutional actors (Lizzi et al., 1996). On the other hand, many local administrations seem to be keen to pass comprehensive traffic plans that are detrimental to a wide array of specific interest groups, which in turn tend to coalesce into a broadbased opposition that the administrations have considerable difficulty in coping with (as in the case of Bologna, and the Modena traffic plan, even though a cooperative approach had been applied), rather than adopting more incremental approaches that would reduce conflict.

One of the basic characteristics of this policy area is the imbalance in terms of influence between organized economic interest groups, that generally oppose restrictive policies, and diffused interests, the victims of traffic pollution. Whereas the former have a strong incentive to mobilize against measures they fear will damage them directly and in the short term, and possess the resources to exert influence effectively, the latter can with difficulty be mobilized to struggle for advantages that are intangible, distributed among a large number of individuals and with benefits that can be felt only in some uncertain future (Edelman, 1964; Olson, 1965). Environmental associations, that have been active on the traffic topic in cities, are hardly able to compete on an equal footing with economic organizations. In such a situation, referendums can represent a source of legitimation in favour of restrictive policies that allow diffused interests to express their preferences on a more equal basis *vis-à-vis* more vocal and better organized interest groups. Referendums have been held in several cities at the request of environmental associations – Bologna (1984), Milan (1985), Florence (1988) – and restrictive policies were supported on such occasions by consistent majorities

Another strategy aimed at creating legitimation can be defined as 'technocratic' in the sense that local authorities resort to experts, preferably foreign (which confers on them further prestige and an image of political neutrality). Bologna, for example, had its traffic plan drafted in 1989 by the German expert Winckler of the University of Munich. The same expert was subsequently asked to work for the city of Florence.

The provision of information concerning levels of pollution through monitoring systems on local newspapers, local radio and TV stations, *ad hoc* bulletins or on monitors and variable message panels (Modena, Parma, Florence) represents another strategy that cities have used to persuade their citizens of the importance of keeping traffic under check. It has, however, proved more difficult for authorities to inform people about the actual implications of the different levels of single, or combined, pollutants, though it is not clear whether warnings from mayors for children and elderly persons to remain indoors on highly polluted days actually favour anti-traffic attitudes, or in the long run induce wearied and resigned fatalism.

It is important to point out that, whatever their strategies, local governments appear to face considerable difficulty in tackling the cultural dimension underlying traffic issues. The private vehicle (car or motorcycle) in this context should not be seen as a mere mechanical means of transport; if the policy debate concerned only the issue of selecting the most effective modes of urban mobility, things would probably be somewhat easier. Along with psychological traits that will not be analysed here, the private vehicle seems to carry with it – especially in Italy – highly individualistic and anti-statist ideological traits (heavily emphasized by advertising campaigns) that are in sharp contrast with the collective management – in whatever form it takes – that the systemic features of traffic require if it is to be at all viable. In this context, an interesting paradox is that, especially in the cities governed by centre-left parties, the rightist parties oppose, not only for reasons of political tactic, both law-and-order measures (respect of basic rules, especially as far as parking is concerned) and economic approaches in the allocation of scarce resources (public space), in contrast with its officially declared market-deregulation ideology.

Implementation and Enforcement

One peculiarity of Italian policy style is represented by serious deficits in the implementation and enforcement of both national and local policies (Giuliani, 1989; Dente et al., 1984). The air pollution and traffic plans, as mentioned previously, have not been drawn up according to specified times. Substantial financial resources have been allocated for parking facilities, mass transport systems and even bicycle paths, but there have been considerable delays in the expenditure of these resources. The poor performance of public bureaucracies and the cumbersome administrative procedures explain a large part of such deficits. The search for consensus, especially among the economic interest groups, typically causes a dilution of the policies in terms of contents or timing, to the extent that they can lose their capability to achieve their original aims.

Even when it comes to simply enforcing basic parking and circulation rules, the lack of personnel accounts for widespread infringement. Local police numbers are low, and those actually on the road are even less, due to the many other duties they are burdened with (and the reluctance of the policemen themselves to stay on the street – partly because of pollution!). Furthermore, the fragmentation of local and state-controlled police forces reduces the effectiveness of enforcement since national forces like the Carabinieri and Polizia seldom care about traffic violations within urban areas. Furthermore, while the number of vehicles continues to grow, the actual number of local traffic policemen in many cities has remained the same due to lack of financial resources. Thus, the dissuasive capability of fines is low.

Such deficits, however, might not be seen as merely unwanted, but rather as the 'Italian way' to compromise in conflict resolution: decisions are taken, but the losing parties can count on the fact that such decisions will not actually be implemented or, if they are, this will occur only with considerable delay and very gradually. For example, restrictive measures on traffic in certain areas are typically not complied with due to the lack of enforcement.

Centre–Periphery Relationships.

Within the policy network, relationships between various levels of government deserve special attention. During the 1980s most innovative measures were initiated at the local level (OECD, 1994, p. 137). The role of the national level at that time was limited to enacting legislation on product standards (for example for vehicles and fuels) as a response to EC Directives. However, as the Environment Ministry started to have reliable financial, legal and operational resources at its disposal at the end of that decade, it became interested in intervening in the field of urban environment and played an active role in stimulating local authorities to act.

Though the municipality is the main policy actor in terms of responsibilities, its actual operational capabilities present a number of difficulties. First, technical offices responsible for traffic management and planning are generally quite weak (by contrast, the case of Modena shows the importance that strengthening such apparatus can have for policy success; Lizzi et al., 1996).

Second, urban sprawl has made apparent the municipal level's inadequacy in managing mobility flows. In the future, the creation of metropolitan authorities provided for by a 1992 national Law should allow government to tackle traffic policy on a wider and more effective basis. So far, however, the process of setting up such new authorities is proceeding with considerable delay.

The third aspect concerns the relationships between the municipality and the other levels of government, regional and central. Until recently, the Italian institutional system has been highly 'centralist'; as a matter of fact, although regions (created in 1970) have important powers in such fields as air pollution (regulation) and public transport (financing), their relevance for urban traffic policy appears extremely limited.

As far as the central State is concerned, it limits the range of municipal policies in two respects. Since 1972, when a thorough reform of taxation was introduced, in practice abolishing local taxes, Italy is the Western nation with the highest degree of fiscal centralization (only 0.7 per cent of taxes are collected by local authorities, compared to 12.7 per cent in the USA and almost 30 per cent in other federal systems; Dente, 1989, p. 155). Thus all resources needed to finance the public system are collected by the central government; quotas are subsequently redistributed to local authorities on the basis of past expenditures. Furthermore, the resources transferred to local government are earmarked for predetermined objectives, as in the case of the National Transport Fund (Ponti and Vittadini, 1990, p. 179), leaving very limited scope for local discretion; local governments cannot hire additional staff such as traffic police or technical personnel because of the national provisions aimed at reducing public expenditure. This situation has begun to be reversed considerably in recent years to meet the growing demand for 'federalism' and local autonomy (and limit public spending through greater delegation of responsibility), but traffic policies cannot be understood without taking into account how deeply it has influenced the capacity of local government to decide its own priorities until now.

A second constraint on municipal policies, especially on the capability of adopting innovative approaches, is represented by State legislation and its (often restrictive and traditionalist) interpretation by courts – especially the Administrative Tribunals (TARs) – as shown by the previously mentioned cases of Sirio and of vehicle emission checks. Generally speaking, municipalities may not adopt measures that are not explicitly provided for by legislation (rather than the reverse principle according to which they would be allowed to do anything that is not explicitly forbidden). The new Traffic Code, for example, though allowing, and even encouraging, local administrations to be more active and innovative, reflects such a philosophy, The above mentioned PUTs are regulated by detailed directives issued by the government in 1995, establishing even such details as how the technical offices responsible for drawing up the plans should be organized and the characteristics of its personnel. The Ministry of Public Works even specifies the experts that municipalities are allowed to use as consultants for drafting the PUTs. Furthermore, the State legislation that local authorities have to refer to is often confused and even contradictory. Regulations in the field of air pollution, for example, comprises some sixteen major pieces of legislation stratified over some thirty years (Desideri, 1996, pp. 15 ff.); nor do the aims and priorities to be pursued emerge clearly from such legislation.

In their relationship with central government, local authorities also suffer from the fragmentation of powers existing in this field among a large number of authorities (Ministries of Environment, Transport, Public Works, Urban areas, national environment agency ANPA) that have different positions and priorities. The case of Sirio is interesting also in this respect: as previously discussed, the Environment Ministry financed it, but it was blocked because the Public Works Ministry had not rubber-stamped it.

For the future, the new legislation, such as the Traffic Code, and the change in the local electoral system give more power to city mayors and make them more independent of the individual members of the town council and pressure groups. These two changes put municipalities in a position to adopt more aggressive traffic policies if they wish to do so. The ball is now back in the municipalities' court.

CONCLUSIONS: ITALIAN POLICY STYLE AND ITS OUTCOMES

The dominating feature of Italy's policies to tackle air pollution caused by traffic emissions is represented by its reactive character, in a twofold sense. On one hand Italian governments have limited themselves to enacting, often with considerable delay, EU legislation concerning fuel characteristics and vehicle emissions standards. The main point, however, is that policy has emerged as a reaction to the perception of the existence of a serious health emergency – the poor quality of air quality in urban areas. As discussed above, the main strategy under this respect has been to call for restrictions to traffic circulation for limited periods of time when thresholds are exceeded. The actual effectiveness of such emergency measures in terms of impact – of their capability to reduce pollution levels – appears from experience to be very limited,[11] at best stabilizing pollution levels rather than reducing them (CSST, 1994), as compared to more structural interventions, such as construction of efficient mass transport systems, that would reduce the number of circulating vehicles and their average mileage. The exemption of catalysed vehicles from the alert and alarm limitations, added to the reduced taxation of unleaded petrol, has encouraged sales of such cars before the legal deadline of 1 January 1993, but it is known that catalysation technology is not effective on the short distances typical of many Italian urban journeys. Furthermore, it should be noted that the Environment Ministry has chosen to focus on peak pollution events, rather than 'ordinary' everyday levels, as if these were acceptable for human health, which is hardly the case (Allegrini, 1994). Restrictions on private traffic have also generated significant levels of conflict at the local level, which municipalities have had considerable difficulties in coping with.

Nevertheless, such emergency policy style has produced one positive effect; that is the promotion of a process of social learning and a change in the actors' belief systems (Sabatier and Jenkins-Smith, 1993). This is also thanks to the considerable attention the whole issue has received from the media. In a situation of relatively weak demand from the public as compared to some of the Northern European countries, the frequent smog alerts and alarms, though hardly effective in reducing pollution, have created a widespread awareness that an actual health problem (rather than a merely environmental one) does exist in large urban areas, producing at least acceptance if not support for alternative traffic policies. In fact, according to a 1992 survey, traffic and air pollution, were considered the most serious environmental issues, because of their direct effects on personal health (Istituto per l'Ambiente, 1994, pp. 36–41 and 51). Several local surveys show that a large majority of the population considers the environmental conditions, and air quality in particular, within urban areas to be bad or very bad. The fact that traffic causes dangerous levels of pollution that constitute a menace to public health, once an argument shared only by a few scientists and the environmentalists, is now accepted as a matter of fact by policy actors and the population in general. Furthermore, traffic has become a hot topic of political debate, as shown by the last electoral campaigns in many cities (CSST, 1994, pp. 9 ff.).

If traffic policy is to become more effective in the future, several conditions closely connected to institution-building will have to be met. First of all, the municipalities need a stronger and more constant support from the 'centre' (whether this is represented by the State or the regions, or more probably both in a federal institutional scenario in the near future); owing to the weakness of each of these levels close inter-governmental cooperation is required. Secondly, the policy network needs to stabilize, and possibly evolve towards a 'community' format where dialogue is regular and less confrontational, in order to plan and implement the long-term strategies that characterize successful policies in this area. Thirdly, in order to achieve stronger consensus, public authorities should realize the fact that they cannot limit their interventions to 'hard' policies (regulation, infrastructures), but that there also exists a 'soft' component represented by individual behaviour. In this regard it is necessary, through information and consensus-building activities, to tackle aspects that public authorities typically shy away from in democratic societies, such as social learning and teaching processes. Finally, and perhaps most importantly, issues in the domain of traffic policy need to be managed by more professional technical bureaucracies, an actor that appears to be still far too weak in Italian administrations. Evidence from a related policy such as industrial air pollution control shows that the most successful administrations in this field were those that first of all staffed their offices with an adequate number of highly competent professionals. The reason for their performance lies not so much in their specific capabilities or wisdom, but rather in the fact that their activity was oriented towards achieving results, not simply towards formal enactment of legal provisions (Dente at al., 1984).

NOTES

* The authors would like to thank Renata Lizzi (DOSP Forli) for the research materials and the useful comments provided.
1. Law no. 615 of 1965 and subsequent Decrees of 1971.
2. Methane gas has become increasingly important in assuring energy supplies (27.3 per cent in 1991) (ISTAT, 1993, p. 192).
3. Measurement of benzene conducted by WWF showed that the quantity of this substance that children are exposed to is equivalent to smoking thirteen cigarettes a day in Rome, eight in Milan.
4. In 1994 the authority responsible for assuring proper competition declared that the use of the term 'green' in advertisements was not admissible (Zolea, 1996, p. 104).
5. The proportion of cars with catalysers has risen from 2.8 per cent in 1990 to 15 per cent in 1993 and to 33 per cent in 1996; it is estimated that it will reach 90 per cent only in the year 2005.
6. It has been estimated by the Automobile Club that in the first year in which the new provision comes into force approximately 19 million vehicles will have to be checked, and that in the following years the number will be about 10 million. By comparison, during the entire period 1991–95, only 6.2 million cars underwent a check-up (*Il Sole 24 Ore*, 11 October 1996).
7. *The Economist*, 12 April 1997, p. 67; *Il Sole 24 Ore*, 4 April 1997. The Minister of Environment stipulated an agreement with the FIAT group on the basis of which the manufacturer will make future investment in the development of vehicles with a lower fuel consumption (especially diesel); the Ministry intends to introduce tax incentives to promote the diffusion of such vehicles. FIAT has started to produce bi-fuel car models that run on both methane and petrol.
8. Taxes on petrol products for transportation were estimated to amount, in 1986, to some 20 000 billion lire (Ponti and Vittadini, 1990, p. 182) and in 1994, to approx. 36 000 billion (Buglione 1996, p. 137).
9. *Il Sole 24 Ore*, 26 April 1997.
10. There is ample evidence that policy success is related to a consensual policy style (Weisäcker 1994, p. 142; Skou Andersen, 1994, p. 56).
11. Between November 1992 and March 1993 in the 15 cities to which the Ripa di Meana Decree applies, there have been 62 days in which alert levels have been registered and 34 when alarm levels have been passed.

5. Shifting tools and shifting meanings in urban traffic policy: the case of Turin

Luigio Bobbio and Alberico Zeppetella

INNER CITY TRAFFIC POLICY IN TURIN

The tools used to control the urban traffic in inner cities are usually a mixture of regulations enforced authoritatively and economic action limiting the supply of traffic related urban facilities (quantity, access or cost). The specific mixture of these two kinds of tool can influence the effectiveness of policy.

There are also two ways to define the traffic problem in inner cities: the first is to stress the issue of air pollution produced by cars; the second is to focus on congestion (the drop in urban efficiency from the standpoint of economic activities and the worsening of urban services for the citizens). Of course, the two are linked; but changes in their perception and relative weight can affect policy-making processes.

The Restricted Traffic Zone

The issue of inner city traffic made its first appearance on the policy agenda of Italian City Councils in the 1980s, mainly on the initiative of the ecologists. The problem was primarily defined as one of pollution and the leading policy tools were regulations to restrict the circulation of private vehicles in central areas. The case of the Turin inner city traffic policy is in this sense a standard one. But in the last few years, things have changed quite radically. The City Council has introduced a policy of parking charges, initially in the central core of the city and progressively in some surrounding areas. This is the first case in Italy of policy change, and is, moreover, successful, thus being taken as a model by some other Italian cities.

The Turin inner city traffic restriction policy (Marzano, forthcoming) first began to be shaped in the mid-1980s by the ecologists. Here, as in many other towns, they promoted debates and published information about the effects of the car traffic on urban air pollution. These effects are of course particularly severe in the central area: the main proposal they elaborated at that time to deal with the problem was to close the city centre to non-residents' cars on weekdays, from early morning to late evening. To support this proposal the ecological associations presented a petition to the City Council in December 1987 requesting that a referendum be held on this matter (note that a referendum of this type can only be advisory in nature). During that winter 15 000 signatures were collected.

The proposal had some effect on the agenda of Turin City Council. The Councillor responsible for traffic prepared a proposal to introduce a 'Restricted Traffic Zone' (RTZ) in the inner city. Compared to the ecologists' requests, the area involved was smaller (155 hectares) and the duration was shorter (only from 7 am to 10 am), but opinions on the matter differed among the political parties in the local government, and in some cases even within the same party. Moreover, shortly after the Council's proposal was drafted, there was a political crisis and a change in City government, so no decision was taken at that time. It was decided, however, to hold an advisory referendum on the problem, concurrently with the elections for the European Parliament, in June 1989. Citizens were asked whether they would agree to measures limiting the circulation of private cars (the question was formulated in very vague terms): 66.4 per cent of electors answered yes.

The result of the referendum speeded up the adoption of the Restricted Traffic Zone, although there was no complete agreement within the majority supporting the City government. In January 1990 the first, small, RTZ was introduced, enclosing the perimeter of the Roman town (about 45 hectares and 7000 inhabitants). The restrictions were in force from 7 am to 5 pm, and the zone had 15 access gates which required 200 local policemen to control them. Following this measure many disputes arose among the various exponents of the City government and to resolve them the Mayor formulated a new proposal. The new RTZ was bigger (155 hectares, about 20 000 inhabitants, 24 access gates) but the time limit was much shorter (initially 7 am to 10 am; but extended to 1 pm in the autumn of the same year).

As we have already said, in the mid-1980s in Italy RTZs were the standard response to inner city traffic problems. The decision-making process in Turin was similar (both from the procedural and the substantive points of view) to that of other big Italian towns (Mazza and Rydin, 1997). Milan first adopted a RTZ in 1984, after a referendum. Other examples of towns adopting RTZs are Bologna and Florence. In Bologna, after a referendum (held in 1984) a large part of the inner city (350 hectares, 25 000 inhabitants) became a RTZ (from 7 am to 8 pm) in June 1986. After pressure from anti-traffic committees and a big public protest meeting, Florence City Council decided, in January 1988, to extend the restrictions on private traffic to a larger area (about 300 hectares) from 7.30 am to 6.30 pm. (These measurers had first been introduced in the 1970s to protect streets and squares of interest to tourists.)

The regulative approach focusing on traffic restrictions seems to be consistent with the national policy on urban air pollution. In the same years, in fact, the Ministry for the Environment issued a decree obliging the mayors of big cities to adopt measures to restrict traffic when the levels of pollution exceeded standard thresholds (banning private cars entirely, or allowing cars with odd or even number plates to circulate on alternate days). However, this way of handling the problem seems to be quite an exception compared to other countries. European local policies on urban air pollution highlight different issues and use different tools: differentiating supply for urban travel; stimulating competition in the collective transport sector, managing and channelling traffic flows; regulating the building of car parks in central areas; protecting residential zones from traffic; charging for parking in overcrowded areas; or introducing systems of road pricing (Bonnel, 1994).

Note that the adoption of RTZs was in all cases the result of pressure from the environmentalists; and one can say that such groups had (and in the main still have) a strong preference for measures that introduce legal obligations, and for the set timing of quantitative thresholds of emission permitted. Moreover, although the importance of distributive and constitutive measures has greatly increased in recent years in the Italian policy for the environment (Lewanski, 1990, 1997), the role of regulative tools is still prevalent, while the use of economic instruments (taxes, tariffs and so on) is almost negligible. On the other hand, environmental issues have only recently started to play a primary role on the agenda of local institutions. They still have limited experience in handling these issues, and there are still no consolidated routines in these fields; there is no rich toolbox, nor standard use of arguments in the justification of policy.

Throughout the late 1970s and early 1980s the urban traffic problem was becoming an increasingly important issue for public opinion, and the city councils of major towns urgently needed to deal with it. In spite of the relative weakness of the environmentalists at that time, it is not therefore so surprising that their proposals to restrict the circulation of private cars in central urban areas had a great influence: it was almost the only idea being offered. Of course, by adopting RTZs, city councils softened the more radical aspects of the environmentalists' ideology.

Perhaps the preference for this type of regulation is linked to the 'scientific' tradition of much of the environmentalist culture. This peculiarity is still deeply rooted, even if it has sometimes been criticized (Ceruti and Testa, 1991; Zeppetella, 1996). Perhaps it is linked to the systemic traditions of ecology as a branch of science. On the other hand, it often happens that both new sciences and new social ideologies have the tendency to search for their own legitimization by appealing to some set of traditional values, some accepted idea or established institution. Radical opinion groups and political organizations often use such appeals as a kind of authoritative argument to legitimate themselves and to widen consent for their proposals. In this sense, the appeal to the authority of science is widely used and seems to be particularly convincing, because of its social neutrality. Because of their universalistic nature, arguments based upon science can be successfully used by public institutions seeking to justify their decisions on the grounds of common interest and social welfare.

The effects of RTZs have, in all the cases we know, often been evaluated negatively. Dissatisfaction is expressed not only by interest groups (especially shopkeepers) who traditionally oppose traffic restrictions, but also by supporters of this kind of measure; even if the environmentalists stress the inadequacy of the measures rather than their failures, emphasizing the need to associate other measures (pedestrianization, increase of public transport, facilities for cyclists and so on) with RTZs. The effectiveness of RTZs is seriously weakened by the increase in the number of permits issued. In the case of Turin, the permits issued soon reached the enormous number of 45 000; 15 000 of which were granted to craftsmen such as electricians and plumbers. It was absolutely impossible to draw up a definite limit for allowing people to circulate within the RTZ; it soon became a sort of slippery slope.

From the point of view of the level of air pollution, there is no evidence that RTZs had a positive effect on air pollution. In the winter of 1993–94 in Turin, for example, there were three or four days of total traffic ban, when the stipulated thresholds for air pollutants had been exceeded.

The Policy Change: City Centre Parking Charges

The turning point in Turin City government traffic policy came in the summer of 1994. The occasion was the presentation of the new Urban Traffic Plan, a new tool for urban policy made possible by the introduction of the new Italian Highway Traffic Code in 1992. One of the main novelties of the regulations as they affected local policy was that they explicitly allowed the City Council to introduce parking charges in all urban areas defined as 'zones of significant importance for urban policies where traffic problems exist'. Before this principle was stated, the prevailing opinion of Italian courts was that parking charges were not legal if there were no personnel watching the cars: this, of course, was not the case with public parking in urban areas.

In 1993 a new city council was elected in Turin. The Councillor responsible for town planning and urban transport decided immediately to take advantage of the opportunity offered by the new legislation. In July 1994 new traffic rules were decided as a 'foretaste' of the Urban Traffic Plan, following an idea of 'planning-in progress', creating dynamic relations between comprehensive policies and specific actions. The main contents of the measures were as follows:

- An hourly parking charge of 1500 Italian lire (about $1) was established for the entire city centre (about 200 hectares) from 7.30 am to 7.30 pm

for all public spaces (about 10 000 places). The price of a monthly ticket for residents was 35 000 lira (about $20).
- The perimeter of the RTZ was slightly reduced, the number of access gates being reduced to 19 and its hours were radically shortened (to 7.30 am to 10 am instead of 7.30 am to 1 pm).
- As complementary measures, some small central areas became pedestrian zones, some protected lanes for buses and tramways were created, and in some streets private and public transport were permitted, but in opposite directions.

These measures sketch a clear policy change. They suggest an explicit trade-off between the introduction of parking charges and the reduction of the RTZ. The proposed transaction was successful enough with shopkeepers (whose opposition was quite muted compared to the protest movement against RTZ). The loudest dissent came from residents; the reduction of the RTZ took away some of their privileges, while the parking charge was a slight but tangible increase in the cost of car ownership for people who were used to parking their cars on the street rather than in a garage. After negotiations, the residents achieved their objective: at the end of September the City government decided to exempt inhabitants from paying parking for six months. This 'provisional' measure later on became permanent.

Parking charges came into force on 17 October 1994. During the following two years the area involved was progressively extended from the core of the city to other semi-central areas. The Urban Traffic Plan was approved in the summer of 1995; it reaffirmed the policy of parking charges, and announced plans to apply them to other areas. The zone progressively reached a size of about 400 hectares, with 20 000 parking spaces (see Figure 5.1). Management of the policy was entrusted to Turin's public transport company (ATM). The City of Turin refunds the operating costs; the net profits are reinvested for works related to traffic and urban mobility. Note that, according to the new Highway Code, non-police personnel may be used to fine those transgressing parking regulations. Thus less costly staff can be employed, radically reducing the costs of enforcement. The policy made a net profit of 10 billion lire in 1995 (about $6 million), out of gross receipts of 16 billion lire (about $10 million).

From the point of view of the effect on traffic and congestion, the parking charge system seems to have been successful. The general opinion is that the situation in the city centre has improved: the congestion is less and, for those really needing to use their private cars in the inner city, it is now easier to circulate and to park. The majority of the citizens appear to evaluate the trade-off as adequate. Moreover, the positive results appear to be confirmed by traffic surveys from survey points shown on Figure 5.2. From June 1994 to June 1996, there was a substantial decrease in overall flows of traffic in central areas,

78   *The politics of improving air quality*

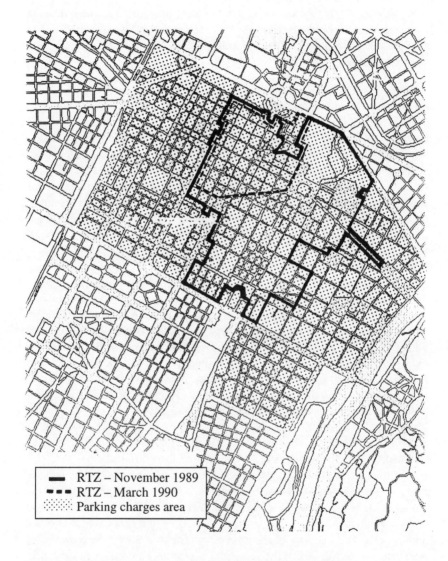

Source: Città di Torino, Assessorato all' assetto urbano, Divisione mobilità

Figure 5.1   Turin: from Restricted Traffic Zone to parking charges

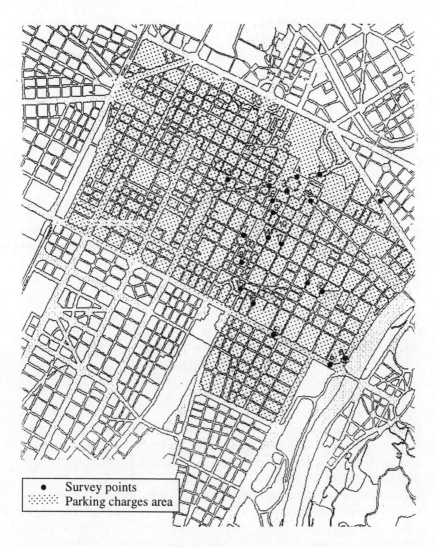

(*Source:* Città di Torino, Assessorato all' assetto urbano, Divisione mobilità

Figure 5.2 Turin: traffic survey points

in spite of some partial increases. This reduction was very marked in the initial period. From November 1994 to June 1996 circulation increased, although it did not come close to the level prior to the introduction of parking charges. This could simply mean that a phase of adjustment followed the first strong reaction to new measures. Furthermore, the decrease in traffic was also important in streets bordering on the parking charges area. This would appear to indicate that there has been a real traffic decrease in the inner city and not merely a redistribution. From the point of view of the effects of parking charges on air pollution, these seem to be positive, even if the measures have been in force for too short a time to say that there is real evidence of a correlation. Anyway, during the winter of 1995–96 the air pollution levels never exceeded the thresholds of acceptability.

POLICY EFFECTIVENESS AND THE MUTUAL GAINS OF COOPERATION

The case of the urban traffic policy change in Turin also suggests some arguments concerning the problem of cooperation in environmental policy. Discussing the ethical implications of the environmental crisis, Vittorio Hösle (1991) stressed that one of the reasons for the difficulty in translating shared values into practical behaviour is the consciousness that the marginal effects of individuals choices are often irrelevant. He quotes Tacitus and his tragic example illustrating such a case. During a battle in a Roman civil war, a soldier realizes that the enemy he has killed is his own father. The news spreads an awareness among the troops that the fight is totally absurd. There is a few moments' hesitation: then, because of the sense of the individual's powerlessness in controlling large-scale events, the fighting starts again.

Urban traffic pollution and problems of congestion can be described in a less shocking but not so different way. The result of each individual's actions is greatly influenced by the behaviour of many other people: so, the choice made by each person appears to have no influence on the general situation. The cost of using private cars in high density inner cities is influenced by the congestion produced by other cars. For each car owner needing to circulate in the urban area, the best case would be to drive while nobody else is using their car.

In such a context, if each individual tries to behave in the standard economic rational manner (trying to maximize his own utility) the result will be very far from maximizing social welfare: each car owner will use his or her car, taking a (marginally worthless) part in generating the tremendous everyday congestion. On the one hand, the number of car users in urban areas is so large that voluntary private negotiations and agreements are quite impossible. On the other, public

Shifting tools and shifting meanings

policies can try to reduce urban traffic, by attempting to force or induce individuals to (partially) assume the cost of pollution control and of cutting back congestion.

Urban Traffic Policy Issues as a Prisoner's Dilemma

A well known way to describe such a situation uses the model of the prisoner's dilemma, where the pay-off of an individual's behaviour strongly depends on other individuals' choices, which are unknown.[1] Normally, the prisoner's dilemma is considered a two-player game. The pay-off scheme resulting from the combination of each player's choice can be that shown in Figure 5.3.

	Column player	
Row player	Cooperation	Defection
Cooperation	R,R	S,T
Defection	T,S	P,P

Notes:
R = Reward for mutual cooperation
S = Sucker's payoff
T = Temptation to defect
P = Punishment for mutual defection
First figures are row player pay-offs: T > R > P > S

Source: Axelrod (1984)

Figure 5.3 The Prisoner's Dilemma

The standard form of the prisoner's dilemma describes a two-player situation. In practice, things are much more complex. The urban traffic issue is a case of a prisoner's dilemma with a large number of players.[2] Let us describe a simple model of a prisoner's dilemma with n players (see Zeppetella, 1996).

Figure 5.4 gives an arbitrary example. The pay-off of each player (deciding either to cooperate or to defect) is a function of the number of cooperating subjects (x is represented here as a percentage of the total number of players, n). The c line is the cost of co-operation; in this example we assume they are constant. In the case of urban traffic policy, these costs can be compulsorily imposed – for example, by banning circulation in the central urban area, as in the case of

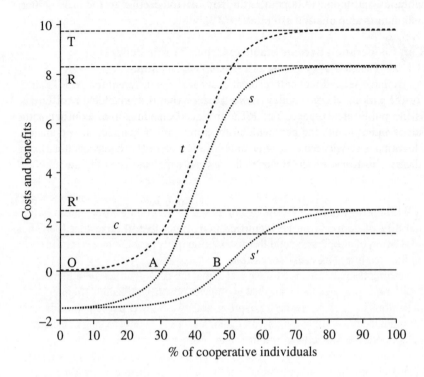

Source: The author.

Figure 5.4 Individual pay-offs of cooperation and defection

RTZs – or by pricing some aspects of use of public goods – as in the parking charges measures. Curves *s* and *s'* illustrate two different hypotheses of pay-off for each 'cooperating' person, while curve *t* shows the pay-off for each 'defecting' individual. To simplify the model, we have assumed that each person, whether 'cooperating' or 'defecting', will have the same costs and benefits.

These curves of individual pay-offs are logarithmic functions, which seems to be likely for many environmental issues, including the case of urban traffic. The marginal effect in terms of a decrease in air pollution decrease will be practically insignificant if the number of those resolving, say, to use public transport or a bicycle, is low: they will increase (slowly at the beginning, then more and more rapidly) until a flex point, where the slope begins to decrease. When 100 per cent of the players less one cooperate, the pay-off for the single subject choosing to defect will have the maximum value of the *t* curve (OT);

when all individuals cooperate, the pay-off for each player will have the maximum value of s and s' (*OR* and *OR'*).

Can Cooperation Become Stable in Urban Traffic Policy?

As we have seen, direct enforcement costs are very different for the RTZ and for the parking charges policy. Nevertheless, what is more radically different is the public acceptance. The RTZ produced an unresolved conflict with shopkeepers, while the parking charges policy never had radical opponents. Negotiations with local interest groups about proposals to introduce parking charges in new areas have generally had positive results; some attempts to organize meetings against the 'oppressive resolutions' met with a complete failure. In other words, it seems that the policy change produced a good measure of willingness to cooperate among citizens. Can the conditions for this relative stability of cooperation be identified? One may hypothesize that both the dimension of cooperation and the distribution of its benefits have an influence on the steadiness of such cooperation.

Starting from the first issues, we notice that the pay-off structure in Figure 5.4 is different compared to the original two-player prisoner's dilemma (Figure 5.3).[3] In the former case, the 'sucker's payoff' is not always lower than the 'punishment for mutual defection'. In our pay-off examples, this is so only as long as the quota of cooperative subjects is smaller than those respectively indicated by *OA* (for s) and *OB* (for s'). That is to say that if the number of cooperative subjects is sufficiently high, the individual return for each of them can, in some cases, be greater than in the case of total defection.[4]

Of course, this does not necessarily mean that such a situation will be perceived as a satisfactory one. If net benefits for cooperation are positive but very low, one might decide to stop cooperating, especially if the benefits for those defecting are comparatively much higher. *Comparative dissatisfaction* can weaken the stability of cooperation. Furthermore, costs and benefits of cooperation may in practice be very different, and their comparison correspondingly difficult. They can concern market and non-market goods, long- and short-term effects; they may be perceived very differently by different subjects, and perception may shift over time. In the case of traffic issues, for example, certain higher costs of using public transport are easily perceived and evaluated by each individual (increase in travel time, risk of delays, dirty vehicles and so on). On the other hand, other effects are less easy to perceive, or are difficult to assess, for example long-term improvement in health due to reduced pollution. If the benefits of cooperation fail to reach a threshold of perception while the costs are clearly visible, the willingness to collaborate with environmental policies can be weak or unstable. In this sense, we may assume that one of the reasons why the policy of Turin RTZ is commonly judged as unsuccessful may be because the area involved was too small to allow the expected benefits for citizens bearing its costs to materialize.

From the point of view of the explicit aim of the policy (reduced air pollution), in fact, the RTZ did not produce tangible results. Even during the phase of its widest application it had no appreciable effect on levels of urban air pollution, forcing the local administration to take the costly measure of stopping traffic throughout the entire city. This is perhaps because the area involved was not big enough to have a significant marginal influence on overall urban air pollution. Moreover, the increasing number of permits and the difficulty of controlling access gates greatly reduced the expected decrease in traffic flows and congestion in the city centre.

The influence of the distributive issue on the steadiness of cooperation is described by the difference between curves s and s'. The first represents the case in which gains arising from cooperation benefit cooperating for those defecting exceed the mere advantage of not paying its costs ($s' < t-c$). The progressive increase of distributive inequality between those defecting and those co-operating make a progressively bigger percentage of co-operation necessary for the latter to reach a positive payoff for the latter ($OB > OA$).

Consider now the total *pay-offs* of cooperation and defection, referring to the same hypotheses as above, and using the same data as in the case of individual pay-offs. Figure 5.5 presents the total returns of both the group of subjects choosing cooperation (curves s and s') and that choosing defection (curve t) for each level of cooperation. The sum of the pay-offs of cooperating and defecting subjects is what we can call the 'social pay-off' (supposing, of course, that the compensation principle can in all cases be applied). In Figure 5.5, social pay-off is represented by the curves $T + s$ and $t + s'$ its maximization respectively corresponds to percentages of cooperation OI and OG.

Note that, if the rate of co-operation is too low, the outcome of cooperation is negative not only for those cooperating but also from a social point of view. In other words, if cooperation does not go beyond a particular dimensional threshold, it has positive effects for a small group of citizens but is not beneficial for the community as a whole.

This is possibly the case of the RTZ in Turin. In fact, the advantages affected a small number of citizens (residents and permit-holders), who could circulate and park in the city centre without conflicting with people not possessing passes. Shopkeepers perceived the RTZ as producing a significant reduction in the accessibility of the city centre and so seriously damaging their own activities (it is not important to discuss here whether they were right or wrong). On the other hand, the expected advantages 'for the city' (the reduction in air pollution and traffic congestion) were too slight (or were so judged). In other words, the burdens for those forced to 'cooperate' were estimated to be too high, even assuming a compensatory assessment criterion.

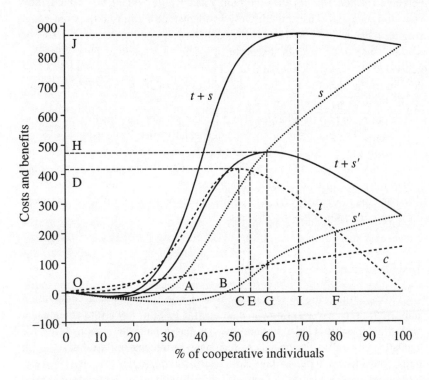

Source: The author.

Figure 5.5 Total pay-offs of cooperation and defection

But things can also be regarded from the point of view of those cooperating and defecting *as groups*. The former reach a positive pay-off when the percentage of cooperation is at least equal to *OA* (for curve *s*) or *OB* (for curve *s'*). The latter will have their best pay-off when the cooperation level reaches the point *OC*. The pay-offs of groups *t* and *s* are the same at *OE*; those of *t* and *s'* at *OF*. Thus, especially when those cooperating and those defecting are identified as permanent groups, we may assume that cooperation will become stable only if the total pay-off for the group is positive. (Of course, this stability will increase when the total pay-off approaches and exceeds the return of the defecting group.) As we can see from observing examples concerning not only individuals' but also groups' pay-offs, the more unequal is the distribution of benefits the smaller is the chance of achieving stability of cooperation.

Perhaps we can now better understand one of the basic reasons for the failure of the Turin RTZ policy. Its benefits, on the one hand, were too small to be perceived as benefits 'for the community as a whole'; on the other hand, they were too unequally distributed. Only residents and permit-holders got some tangible advantage. Thus, the only logical behaviour for those compelled to 'co-operate' was to try to join the group of 'defecting' people. This behaviour generated an irresistible pressure to get permits; but the increase in these also decreased the benefits for those allowed to circulate in the RTZ: a vicious circle that was very hard to break.

The policy of parking charges also produces greater benefits for residents in areas where the regulations are in force (especially after the decision to exempt them from paying for parking near their homes). Nevertheless, decreasing traffic congestion, the improved quality of inner city life, the possibility of finding a parking place in central areas when needed, the reduction of total traffic flows and the resulting decrease in polluting emissions are positive results, even for people living in other urban areas.

THE INSTITUTIONAL ARRANGEMENTS

We have seen that the policy change was able to favour greater cooperation among car users. We want now to examine the institutional arrangements that made this possible. Under Italian law, the mayor has sole responsibility for regulating city traffic. Provincial and regional governments deal exclusively with traffic outside what Italian law calls *centri abitati* (cities, towns, villages). Moreover, they have no interest in getting involved in such controversial matters, preferring to leave it to the city.

The policy change in Turin was favoured by a change in national legislation. The new Highway Traffic Code of 1992 allowed city mayors to introduce charges in unmanned parking areas, while under the previous law it was unclear whether the cities had the right to impose such payments. Some courts had denied this possibility as, in their opinion, a true exchange was lacking; the charge was to be considered as a tax, and as such was not due. Parking charges were not forbidden, but deciding to levy them was risky for city councils. By clarifying this point, the new law gave an opportunity to Italian cities, but only the City of Turin chose to seize this opportunity, two years after the law was enacted.

Coming now to the processes that took place within the local context, we shall compare the two policies through four different features: who the initiators were, and how the solution was sought and found; the groups that supported or opposed the decision, and the interaction patterns which developed among them; who controlled the process of implementation and development of the policy; and how problems were defined.

The Initiators

Banning cars from Italian inner cities was one of the first and most important successes of the newly born Green Party (the other success being the 1987 referendum that halted the national plan for nuclear power plants). In the period 1980–85 the Greens had for the first time entered the national parliament as well as several local councils, and thus were in search of initiatives that could strengthen their public role. Private traffic in urban areas was immediately seen as a key issue. At that time, Italian cities appeared to be backward compared to other European cities: there were few pedestrian areas, no cycle paths, and on-street all parking was free of charge. And the traffic was chaotic in the old streets and *piazze* of the inner cities. The Greens then proposed a radical approach: as private cars were responsible for most of the air pollution, they should be banned from the 'historic centres'. Their model was the 'city without cars'; their preference for a regulatory policy style. They started their campaign in Milan in 1984, from where it was extended to other cities. The pattern was always the same: they called for a local referendum, they won it and then claimed that the city council had to implement the 'will of the people'. It may be asked why this radical proposal was so easily accepted by voters and local government. There are at least three reasons: first, the awareness of being backward provided pressure for a big step forward; second, there were no clear alternatives, as the local policymakers had not devised any alternative solution; third, the local policy-makers knew that something had to be done and saw the result of the referendum as an opportunity to act without taking full responsibility for their action. The overall outcome was paradoxical: while other European cities had proceeded in this matter through incremental changes, Italian cities decided abruptly to overtake them through a radical measure that was not in tune with the traditional style (incrementalism, bargaining and conflict avoidance) of Italian policy-making.

The city of Turin was a latecomer. The referendum was held in 1989 and the RTZ was introduced in 1990. In this case, the policy was shaped by the Greens through mere imitation. There was no discussion about what solution to provide in response to what problem; the RTZ appeared to be an obvious and natural solution, as it had already been adopted by the other major Italian cities. Should Turin lag behind Milan, Bologna, and Florence?

The story of the second policy (parking charges) was completely different. The idea was worked out at a technical level by the traffic experts involved in preparing the Urban Traffic Plan (in Table 5.2 this change is labelled 'innovation'). Of course the experts were strongly influenced by the experience of other European cities, and imitation had an important role. But in the case of Italy, this solution could not be introduced without breaking away from the previous approach. It required a complete change of approach to traffic management. An important role in initiating the new policy was played by the municipal administration elected in 1993, a year in which, after the crisis of Italian political parties, a new electoral law had introduced the direct election of the mayor. The effect of electoral reform was to reduce the influence of political parties and increase the role of people elected from outside the world of politics. Valentino Castellani, Professor of Engineering at the Polytechnic of Turin, with no previous experience in politics, was elected mayor, at the head of a centre-left coalition. The urban planner, Franco Corsico, a professor at the same university, was appointed councillor in charge of traffic and planning. Corsico was ready to resume the experts' work and to launch their ideas about parking charges. Their move was not introduced by a public campaign; people only became aware of the policy change when it was introduced. There was a sharp difference between the two start-ups. In the first case, the issue was raised by a political group, with strong emphasis on environmental values and change; in the second case, it was defined by a close circle of experts as a technical tool to improve traffic conditions.

Supporters and Opposition

The RTZ proposal met the immediate opposition of two different groups: the retailers of the city centre and the people living outside the 'prohibited area'. Both groups had an important influence on the policy implementation. Just a few weeks after the introduction of the RTZ, the inner-city retailers complained that their sales had dropped sharply, as people were not permitted to reach the city centre by car. An *ad hoc* committee of retailers organized a public campaign for an 'open inner city', stressing that closing the city centre meant killing it. Traditionally, retailers constitute an influential lobby in Italian cities and in Turin, their opposition had the effect of dividing local policy-makers, who themselves were not very enthusiastic about implementing the result of the referendum. In fact, they began by restricting traffic in a tiny area and after extending it they shortened the duration of the restriction from fourteen hours to five hours a day.

The highest price for the RTZ was paid by people living outside the borders of the prohibited zone, as it prevented them from driving from the suburbs to the inner city. Through different associations, they applied strong pressure on the city administration to extend the categories of permit-holders; and the administration was often obliged to accept their demands, thus weakening the policy. Thus, acceptance of this measure was quite limited, the only beneficiaries being the people who were resident in the inner city. While the city administration was ready to adapt the policy to silence protests from its opponents, no real negotiation took place between the two sides. The latter continued to feel completely dissatisfied by the measures adopted, as they were playing a zero-sum game.

When the policy changed, a new and more favourable set of interests were involved. The inner-city retailers weakened their opposition. At first they perceived the parking charges as the lesser of two evils, but they soon realized that the new rules were improving their situation: people were more willing to shop in the centre as it became easier to find a parking place. In this case too, people living outside the paying zone applied pressures for free parking, but it was easier for the city government to resist: making people pay to park was seen as a less serious measure than denying them the right to drive into the city. Problems of equity were (weakly) raised by the left-wing opposition, but they were easily refuted, as the charge was not so high as to prevent low income groups from driving into the centre in their own cars and parking for a few hours.

Within the political arena, the councillors from the Green Party were not enthusiastic about the new solution, because it appeared too smooth. They accepted it, as they were part of the city government, but they claimed that the RTZ should not have been abolished. Eventually a compromise was reached: the RTZ was retained, but within a much smaller perimeter and with a shorter period of time (8 am to 10 am). Such an RTZ still exists, but has very little effect, no control being in place at the gates.

Initially, the strongest opposition to the policy change came from the residents in the centre. The first proposal of the city administrators was to make residents pay to park their cars, although in the form of a lump sum of 35 000 lire per month (about $20). This measure was forcibly rejected by a newly constituted citizens' committee, who objected: 'We can't be obliged to pay to go home!'. There were demonstrations in front of the Town Hall, but the city administrators were wise enough to avoid a confrontation and to negotiate free parking for residents all day, though within small areas. This agreement was the condition whereby the policy was able to take off and spread, since the newly involved people had nothing more to fear.

Policy Implementation and Development

The different configuration of interests does not in itself explain the difference in effectiveness of the two policies. In the first case, no actor played a pivotal role: the proponents (the Greens) were by that time out of the city government, while the policy-makers in charge were somewhat reluctant to commit themselves too deeply. In the second case, the initiator was the councillor for traffic, who was able to guarantee continuous control over the implementation of the policy.

Moreover, the two policies displayed different degrees of flexibility. The RTZ measure was, by its very nature, quite rigid. The city administrators tried to move through incremental steps: they started with a very small area and gradually extended it; they altered the times of the ban to suit the public mood. Objectively, however, they did not have much room for manoeuvre: a restricted traffic zone cannot be extended beyond certain limits, and it creates problems of congestion in neighbouring areas. The vehicles that were not allowed to enter tended to crowd around the borders in search of a parking place as close as possible to the gates of the RTZ. Congestion disappeared inside the zone, but was exported to just outside it; the problem had been shifted rather than solved.

The situation changed completely with the adoption of the parking-charge policy. The new tool appeared to lend itself to incremental changes. When the policy was introduced in 1994, it affected only one area of the city centre – corresponding approximately to the old RTZ – and new areas have since been gradually added. In this case, the congestion exported to neighbouring areas had a positive effect on the development of the policy, as people living in those areas had a definite interest in the introduction of paid parking near their homes, in order to shift the congestion further on. The parking-charge policy was self-sustaining. Moreover, at each step the administrators were ready to negotiate with the interests involved, in order to adjust the measures (charges, allocation of blue lines) to the needs and problems of the neighbourhood concerned. Currently each area has slightly different parking arrangements.

Both policies entailed heavy enforcement costs. The RTZ required continuous control at the zone's gates to stop unauthorized vehicles from entering and the parking-charge policy required systematic control on parked cars to make sure that the charges had been paid. But in the latter case, costs were covered by the receipts from the parking charges, and a surplus was created that allowed new public investments in the traffic policy. While in the former case all costs were at the expense of the municipal budget, in the latter a virtuous circle was generated: both the cost of enforcement and the investments for expanding the policy were paid for by the policy-takers, that is, car drivers.

Moreover, in the case of the RTZ, control was exercised by 'normal' policemen who had to be moved from other tasks: one of the reasons for failure is that not all the gates were manned all the time, and some 'illegal' vehicles were able to cross the forbidden line. On the other hand, the supervision of the parking spaces was assigned to a public agency which hired special personnel for this purpose, and control was thus more satisfactory.

The keys to the success of the parking-charge policy are thus twofold: a continuous search for consensus through negotiations and incremental changes; and a higher degree of institutionalization – the agency charged with the management (and the personnel they hired for this purpose) had an interest in retaining and expanding the policy.

Defining the Problems

The two policies were presented as responses to two different problems. The RTZ was conceived as a measure to tackle the problem of pollution, and was strongly supported by the ecologists. The policy of parking charges was conceived as a technical tool to tackle the problem of congestion and to offer car drivers a better chance of finding a parking space.

The way the problems were defined had a significant impact on the feasibility of the policies. In principle, it is obvious that urban traffic causes both pollution and congestion, but stressing either feature may make a critical difference. Arguments based on clean air tend to be perceived as ideological and conflicting with urban mobility: a sort of punishment for car drivers. Arguments based on congestion are related to the aim of improving urban efficiency – strongly rooted in common sense. Pollution is less visible than congestion; indicators of clean air can be detected only by experts and often appear to be questionable and reduction of pollution is long-term in effect. On the other hand, the goal of reducing congestion is easily understood, and the results of such policy are clearly visible in people's everyday experience. For an individual it is easier to calculate how much time he or she has saved driving from the suburbs to the centre or searching for a parking place, than to judge whether the condition of the air has got better or worse. Even if all citizens were seriously committed to environmental policies (and this is not the case in Italy) they would have some difficulty evaluating their benefits.

Policies are more feasible if they are coupled with problems that are generally perceived as serious (Kingdon, 1984)[5] and if their outcomes can be easily evaluated in the short term. The case of traffic policy in Turin shows that the environment can be improved as the indirect result of a policy primarily concerning other issues.

NOTES

1. The prisoner's dilemma is a typical non-zero sum game, while neoclassical environmental economics normally supposes that environmental costs and benefits are a zero-sum game (see the standard comparison between marginal cost of pollution and marginal cost of pollution control). These costs can be differently distributed, but they are, by definition, the same for each level of pollution. Nevertheless, problems concerning environmental issues are frequently non-zero sum games. This is why many environmental resources are non-market goods, and they are often the source of positive and negative externalities. In environmental matters, many costs and benefits are not exchanged in an explicit way. On the other hand, environmental externalities

often involve all the people living in a common space (for example urban air pollution); in such situations, actions modifying present conditions can produce mutual gains (or losses).
2. In discussing the use of the prisoner's dilemma to examine urban inner-city traffic policy, we use the terms 'costs', 'benefits' and 'pay-off' in a wider sense, including market and non-market aspects, and economic and environmental effects. Clearly we cannot here go into the problem of monetary evaluation of externalities and 'intangible' goods.
3. The structure of the pay-off matrix is the same in other versions of the prisoner's dilemma, while figures are different (see for example, Richardson, 1978; Kraines and Kraines, 1993).
4. We can imagine examples in which the cost of cooperation always exceeds net benefits for the cooperating individual.
5. Kingdon (1995, p.173) shows that urban mass transit policy in the USA was alternatively conceived as a response to congestion, to pollution and to the energy problem, with different outcomes.

6. Changing definitions and networks in clean air policy in France

Corinne Larrue and C.A. Vlassopoulou[*]

INTRODUCTION

As indicated by the title, our intention in this chapter is to study clean air policy in France by establishing a correlation between two variables: problem definition, and the actors involved in clean air policy. We will try to show that the dominant perception of air pollution reflects a specific configuration of the actors involved. In that sense the redefinition of the policy problem must necessitate a restructuring of the policy network, either by changing the positions of the present actors, or by the emergence of new ones.[1]

Social problem theorists have focused for a long time on the process of redefining of social problems (Spector and Kitsuse, 1987). This has not been the case in policy analysis, where the question of the distinction between objective conditions and the definition of these conditions as public problems has been raised only recently. Edelman (1991, p. 5) recognizes that 'problem definition determines authority, social status and financial resources for some policy actors and refuses them for others', while Rochefort and Cobb (1994, p. 5) accept that 'issue definition and redefinition can serve as tools used by opposing sides to gain advantage'.

Our approach, considering issue definition as a central element in the policy process, follows this train of thought. Policy actors do not take public problems for granted; they negotiate not only about solutions but about problem definition as well. In fact, the two elements of negotiation coincide because the solutions advanced depend on the prevailing perception of the policy problem. This must not be seen as a simple policy sequence, as Jones (1977, p. 10) has suggested. Each participant in the policy structure defends his own perception of the problem, in accordance with his own values and interests, throughout the policy process. Therefore a redistribution of power within the policy network would mean a change in the dominant precepts concerned with specific air pollution problems. Next we will examine the recent consideration of air pollution as a car traffic problem. This redefinition led in 1996 to a new clean air law and the modification of the composition of the policy network.

THE CAPTURE OF CLEAN AIR POLICY BY INDUSTRIAL INTERESTS

Looking from a historical perspective it is possible to identify two different policy networks concerned, in the beginning, with two distinct air pollution problems: the industrial network dealing with industrial pollution; and the automobile network dealing with car pollution. In the case of France, not only had these networks never been in contact prior to the publication of the first clean air law in 1961, but also the two problems were characterized by a very different degree of politicization. The visibility and the harmfulness of industrial locations provoked, from the very beginning of the industrial era, the hostility of the surrounding area. During that period hygiene was a major governmental priority and health specialists were invited to collaborate closely with public authorities.

Before Pasteur's discoveries public hygiene theory was based on the hypothesis of air quality: the spreading of 'miasma' was considered to be transmitted through air, and health specialists acquired their expertise through the analysis of atmospheric composition. According to these perceptions, industrial pollution was integrated in the pre-existing theory of sanitary living conditions and defined as a health problem; its regulation was shared between health specialists and the Ministry of Industry. Since then, for over a century a conflict of interests existed; health specialists were trying to reinforce regulations, while the industrial lobby was looking to weaken them. Car pollution, on the other hand, has never progressed beyond the frontiers of a tightly closed network comprising the car industry and the Ministry of Transport. Automobile industrialists were powerful enough to keep vehicle pollution out of any public debate until the 1960s. Furthermore, the dominant perception of the car as a symbol of personal and social success has hindered any focus of attention against this source of pollution. From 1903 the government instructed an extra-parliamentary committee composed of automobile experts and industrialists to define all technical questions concerning cars, including that of pollution (*Official Journal*, 1903). Therefore, no regulation of car emissions occurred throughout that period.

These discrete networks were restructured and came in contact for the first time when air pollution was defined in the international agenda of the 1950s as a complex problem resulting from different pollution sources. This occasion served in both networks as an opportunity to control the development of clean air policy. Thus new legislation does not seem to have radically changed the way in which those two pollution sources were regulated in the past.

The Construction of a Network Prior to the Definition of the Clean Air Problem

According to our hypothesis, the definition of the air pollution problem at the beginning of the 1960s would provide the restructuring of existing policy networks concerned with different pollution sources. In fact, the main industries who were accused of polluting the atmosphere and the vehicle manufacturers were mobilized by creating a common arena of negotiation in order to participate in the new clean air policy.

This was achieved by setting up structures of expertise which permitted them to become the privileged participators in dialogue with the public authorities. In 1958 the French Electric Company (EDF) created the Association for Air Pollution Prevention (APPA). Knowing that the Ministry of Health was still dominant in this policy issue (the setting up of the Ministry of the Environment in 1971 marked the transfer of policy responsibility from one ministry to the other), EDF tried to gain its trust. With the support of the Ministry of Health and the National Institute of Health, the EDF used APPA in order to finance research in the air pollution field, to obtain the measurement of emissions from its own plants and to ensure that an expert would be able to represent its interests in the development of new regulations.

Two years later the French Combustion and Energy Institute created the Technical Action Committee for Air Pollution (CATPA) in order to represent all industrial branches in the clean air policy negotiations. The Inter-professional Technical Centre for Air Pollution (CITEPA) was a small structure within the Technical Action Committee. Its role was to supply industrialists and administrations (principally the Ministry of Industry) with technical information for the development of new legislation. In fact, the 1961 Clean Air Law and the decrees for its application were essentially based on propositions presented by this committee to the government. During the same period the automobile industry acted in a similar way by creating a Committee for Action Against the Contamination of the Atmosphere (CAPA), entrusted with representing this specific branch of industry in clean air policy negotiations. Inside CAPA a small technical group, the Technical Union for Automobiles, Cycles and Motorcycles (UTAC), was also instituted. As a unit of automobile experts, UTAC was entrusted by the industrialists with defining the economic feasibility of clean air policy measures for the automobile industry. Simultaneously, the Ministry of Transport confirmed this union as the official expert on the technical control of vehicles. Thus UTAC became the clean air policy adviser for both the industrialists and the administration.

Designated the coordinator for the development of a clean air policy, the Minister of Health could not ignore the presence of these structures which represented strong economic interests and whose prior agreement was necessary for the success of his activities. During his parliamentary speech of 18 May 1961, he announced his intention of collaborating closely with them, while the President of the Technical Action Committee was redefining each actor's area of competence: 'our role consists of ensuring a permanent relation with governmental institutions and especially with health specialists entrusted with determining the causes and effects of air pollution ... the committee will define, case by case, what it is possible to do and will give to governmental decision a tangible content' (Technical Action Committee Report 1960–61).

From that moment on, the strengthening of industrialists marked the progressive weakening of health specialists in the new clean air policy network. Furthermore, this evolution permitted the reconfiguration of the industrial network composed, from then on, of the close-knit coalition of industrialists and the Ministry of Industry. By adapting to the new policy context, the previously independent networks combined in a common arena in order to define the new policy problem. A closer look at the negotiations that took place during the development of the 1961 Clean Air Law, and at the content of this law, shows that air pollution was made compatible with the interests of those who dominated its definition.

Definition of the Clean Air Problem as a Reflection of the New Policy Network

In 1961 air pollution was defined in the international agenda as a problem resulting from three major pollution sources: industry, motorcars and central heating. Under these circumstances, the coalition of the industrial and automobile structures could only influence the definition of the degree of responsibility incumbent on each source of pollution. It is not surprising to observe that, even if the Clean Air Law no. 61–842 of 1961 refers to all pollution sources, industrial and automobile pollution is presented in a way that minimizes its importance and compromises its future regulation. By the development of expertise in this domain, the policy actors used the formulation of clean air policy as an opportunity to confirm their control of these air pollution sources. This was achieved firstly by the development of a specific discourse mechanism and secondly by the administrative organization of clean air policy.

The 1961 law was presented as the result of the previous legislation relating to the location of industrial plants. Using this precedent, the representatives of CATPA affirmed during negotiations for the new law that 'the existing legislation has permitted the combating of most of the air pollution problems' (CATPA Information Bulletin, 1960/1961). The same argument was adopted by the speaker of the Commission for Economic Affairs and Planning in the parliament who remarked that 'in our opinion, we have talked a lot in the National

Assembly about the responsibility for industrial emissions, while other sources are more significant' (Official Journal, 1961).

In the case of automobile pollution, there was no previous legislation that could be used to affirm that the problem was already on its way to being resolved. Thus a different type of debate was held to distract attention from this pollution source. The speaker of the Commission for Economic Affairs and Planning contended that automobile pollution is essentially part of a larger problem of urban concentration so could not be treated independently. A deputy affirmed that this kind of pollution was linked to the improvement in the standard of living and for that reason it was not possible to limit it (Official Journal, 1961).

As far as automobile industrialists were concerned, they were trying to keep this problem out of the political arena so as to continue regulating it through the tightly closed automobile network. Thus they did not intervene openly in the debate. The same method was not developed in the case of central heating, for which a consensus appeared, between those actors, about the necessity of regulating this new area of pollution.

Furthermore, by putting pressure on the legislative procedure, the industrial and automobile networks obtained concrete guarantees concerning the application of clean air policy in the 1961 law. Firstly, this law confirmed that the distribution of responsibilities between ministries remained the same: the Ministry of Industry was responsible for industrial pollution and the previous involvement of the Ministry of Health became progressively weaker; the Ministry of Transport continued to monopolize the problem of automobile pollution. Secondly, the control of air pollution also remained the same: the control of industrial emissions was organized by the pre-existing industrial legislation under the responsibility of the powerful Corps des Mines (the prestigious civil service body of mining engineers responsible for industrial development) (see below), while the control of automobile emissions was defined by the road legislation under the responsibility of the Ministry of Transport.

In other words, the 1961 law did not replace previous industrial and automobile legislation but reinforced the fragmentation of the clean air legislation. From then on, any regulation concerning industrial emissions was adopted in reference to the pre-existing industrial legislative framework (the only exception being the decree of 11 August which excluded the most polluting industrial plant). Similarly, the regulation of automobile emissions was organized through the road legislation. The main source of pollution recognized and regulated by the decrees of the 1961 law was central heating.

The creation in 1971 of the Ministry of the Environment does not seem to have influenced the control of clean air policy by the interests it was supposed to regulate. As far as industrial pollution was concerned, the transfer of the Department of Industrial Environment and its personnel (Service de l'environnement industrial) from the Ministry of Industry to the new Ministry, guaranteed its control by the *Corps des Mines*. This transfer brought about the creation of the Office for the Prevention of Pollution and Harmful Substances (Direction des pollutions et nuisances) which was also assigned to the *Corps des Mines*. In 1973 the Department of Air Pollution (Service des problèmes de l'atmosphère) was created within this Office. Management of this was given to an engineer in the *Corps des Mines* who was previously Head of Department at the Industrial Environment Office.[2] Automobile pollution continued to be regulated by the closed automobile network without any particular transfer of competencies or structures to the new ministry.

Therefore it appears that, by this coalition, industrialists succeeded, first, in defining the clean air problem by minimizing the degree of responsibility incumbent on automobile and industrial pollution and by emphasizing central heating pollution; second, in fragmenting clean air policy regulation in the same way as before 1961.

Nevertheless as the international agenda of the 1960s influenced the definition of air pollution at national level as a complex problem resulting from different pollution sources, so a new conjuncture will permit the withdrawal of automobile pollution from obscurity by redefining air pollution essentially as a car traffic problem. According to the initial hypothesis, this will provide a reassessment of the previous clean air policy network as well as the developement of a new law to give shape to this restructuring.

WHEN AUTOMOBILE POLLUTION BEGINS TO REPLACE INDUSTRIAL POLLUTION: TOWARDS A NEW POLICY NETWORK?

The second part of this chapter is devoted to the analysis of an on-going change in the French air policy. As a matter of fact, as a consequence of the changes explained above, the implementation of air pollution policy during the 1970s and 1980s and up to the beginning of the 1990s was concentrated on industrial air pollution. But recently a change has occurred in the problem definition of the policy, at the central level, and for most actors involved in the policy network. This is a change that tends to emphasize car exhaust as the main source of pollution instead of industrial emissions. It will change the clean air policy network configuration, at local level in a near future.

The following paragraphs will outline the main characteristics of the policy implemented during the past fifteen to twenty years, and then present the main factors which influenced and characterized the change of policy problem perception and policy programme.

French Clean Air Policy in the 1970s and 1980s: a Juxtaposition of Two Closed Realms

The dominant position of industrialists as described above influenced the form of implementation of the French air policy during the 1970s and 1980s. Until the beginning of the 1990s, the main characteristics of the air policy implementation were as follows:

1. A localized policy – one which is devoted to tackling local emissions of pollution in order to improve local air quality.
2. An ambient air quality policy based on the 1961 law and its 1974 decree, and also on the industrial pollution control law of 1976 (which reforms the 1917 law) and its 1977 decree, the enactments of which have been described above. On these legal bases, three main instruments allow the reduction of air pollution: the limitation of individual industrial emissions and/or the improvement of pollutant dispersion through the permit given to 'classified' plants under the industrial pollution control law; the creation of special zones for air protection, located mainly in urban areas and within which the quality of the fuel consumed is controlled; and alert on zones, where restrictions to fuel consumption are imposed, on major SO_2 emitters when a pre-defined threshold is exceeded.
3. The main target groups of this policy were industrialists; the main pollutants were SO_2 and smoke; the main pollution problem was a winter one; and the main administration responsible for this policy was the Ministry of the Environment but also the Ministry of Industry and its decentralized bodies the Regional Office for Industry, Research and Environment, DRIRE (Directions régionales de l'industrie de la recherche et de l'environnement) within which the dominant body is the Corps des Mines. The organization of these decentralized bodies has changed over the period. At the creation of the Ministry of the Environment, there were fourteen *arrondissement minéralogiques* and five *circonscriptions electriques*, within which was located the 'mining service' charged with industrial pollution control; in 1976 nineteen interdepartmental offices for industry (DII) were created, which in 1981 became twenty-two (DRIR; DRIRE since 1991).

On the national level, a closed arena was set up during the 1960s which lasted until the mid-1980s and included the Ministry of the Environment (especially the SEI (Service environnement industriel)), the Ministry of Industry (especially the office in charge of fuel control), and the representatives of industrialists, organized through CITEPA (see above) which is a technical representative of industrial polluters. The main regulations regarding air pollution and industrial emissions were discussed within this closed network. The inclusion of the

Ministry of Industry has two main explanations. On one hand, the Ministry of Industry is the principal actor responsible for the regulation of fuel composition and succeeded, through different regulations, and in conjunction with the EU related directives, in reducing the sulphur and lead contents of fuels. On the other hand, the ministry, (especially the General Directorate of Industry) acted as a defender of industrial interests.

The functioning of such an arena was, however, also dependent on local policy networks, which acted in partnership with this national arena. Indeed, at the local and regional levels, equivalent arenas have been set up which include mainly industrialists and decentralized administration of industry (Ministry of Industry regional representative; DRIRE), local governments in association with bodies concerned with public health and environment issues (such as local APPA branches), in some cities such as Lille, Strasbourg or Bordeaux.

The outcome of such forums was the creation of local networks for monitoring air pollution in those cities where industrial emission was considered dangerous. These networks focused on SO_2 and dust immission[3], and partly on CO and NOx emission. They were financed by different partners: central government, local governments, and industrialists through non-profit organizations. These management structures acted as real arenas where air pollution problems could be discussed with all the interests involved, and where the implementation of different means of intervention might be 'negotiated' between the main protagonists. Then, for instance, on the basis of these networks, alert zones were defined in some especially polluted areas, the threshold of which is based on SO_2 levels (that is on industrial emissions) and negotiated through local associations in charge of the management of the monitoring network.[4]

The results of such a policy are not unsatisfactory.[5] Air pollution stemming from industry and from domestic heating has obviously decreased. From 1980 to 1990 the industrial SO_2 emissions were divided by three, (this reduction being mainly due to changing energy supply choices – towards nuclear energy for electricity producers, towards gas and electric sources of energy for individuals and for industrial plants). As a consequence, the air quality of the monitored industrial and urban areas, shows an obvious improvement: in most monitored cities the dust and SO_2 emissions have decreased even if the trend has slowed during the last two years. Such a conclusion is also true for CO and lead emission and 'immission'.

However, in relation to NOx emission and NO_2 and ozone 'immission' the situation is not satisfactory as the reduction of industrial emissions has been offset by the increasing of emissions from transportation. Up to now, the transportation sector has represented about two-thirds of NOx emissions in France. In a number of cities the guiding value for NO_2 set in the EEC directive 85-203 has been exceeded several times (twice in Strasbourg, Paris and Toulouse, and four times in Marseille in 1992, for instance). The absence of concern for air pollution in transport policy largely explains this development.

Transport problems and solutions
As a matter of fact, during this period, car pollution was covered by means of vehicle emission standards imposed on car manufacturers. The drafting of those emission standards fell within the orbit of the Ministry for Transport, even though they were deeply influenced by concerns at international and further EEC levels.[6] For instance, a representative of this ministry took part in the negotiations at EU level, about the revision of the 1970s Directive no. 70/220, which led, at ministerial level to the so-called 'Luxembourg compromise' and made 'clean' cars compulsory. Although liaison with car producers is officially under the jurisdiction of the Ministry of Industry, a close relationship exists between car producers and the Ministry of Transport (which appears to be the most important partner of the car industry, rather than the Ministry of Industry). The Ministry of Transport, the main technical 'corps' of which is the Corps des Ponts et Chaussées, is also responsible for the regulations governing drivers' behaviour, as well as for road planning and building.

The ministry is very well related to the territory through its decentralized bodies the Regional Office for Infrastructure (Direction regionale de l'equipement or DRE) and the Departmental Service of Infrastructures (Directions départementales de l'Equipement or DDE). It is assisted on the technical side by a number of technical and scientific bodies, with the appropriate expertise to tackle problems with transport: for example, CETE (Centre d'études techniques de l'equipement); INRETS, (l'Institut national de recherche et d'étude des transports et de leur sécurité); or the CERTU (centre d'études sur les réseaux, les transport l'urbanisme et les constructions publiques). All of these bodies are under the overall control of the Corps des Ponts et Chaussées.

As for vehicle emissions, the problem was considered by the Ministry of Transport to be under control, since the emission standards imposed on new cars progressively reduce the emissions of individual vehicles. Regulation of driver behaviour was not considered necessary, because the problem was regarded as a technical one, requiring technological know-how and negotiation with the automobile industries. No attention was given to reducing car traffic, or to controlling the behaviour of car drivers in relation to their air pollutant emissions. As a consequence, within this context, the Ministry of the Environment was more or less excluded from the field. However, the problem of urban mobility was considered by the 'transport' actors simply as a congestion problem.

These actors traditionally tackled the problems of car traffic without any consideration of air pollution, and with an approach based principally on equipment. Since World War II, the solution to congestion was mainly sought by building new roads, planning new towns, and, in large urban areas, building new public transportation infrastructures. More recently, faced with the growing importance of congestion, the main approach of administration and cities has been to improve the flow of traffic by new advertising systems (by indicating the time for a given journey route, as in Paris for instance), but up to now, restriction of traffic as such has been prohibited, except in some cities like Strasbourg, (although this has been more to improve the quality of life than the quality of the air). Moreover, the development of individual mobility was – and still is – considered desirable. Urban mobility was associated with social mobility, so access to personal motorized mobility has became a social objective.

The 1982 national transportation law (LOTI), which regulated the organization of competencies in the field of transportation, tends to 'rationalize' urban mobility, through the formulation of urban mobility plans (PDU: Plan de déplacements urbains) which were not compulsory until the recent clean air law. The objective of such plans was to balance the distribution between private and public transportation, in order to tackle congestion problems, but also to guarantee and strengthen the right to be transported especially by public transport. The air pollution argument did not really exist at that time so, until recently, reduction of air pollution was not included as an objective in the formulation of PDUs.

As you can see, two different policies formulated and implemented by two different corps of civil engineers with two different philosophies, are the main explanation of the delay in including mobile sources of pollution in air pollution policy. Each of these arenas had a different air pollution problem perception related to the composition of the arena. Furthermore, until recently, car policy hardly targeted clean air policy, except in the field of clean cars which was imposed from EEC level and not really targeted to urban air pollution (see above).

Such a juxtaposition of arenas is mostly obvious at the local level. Some local authorities which have important legal powers to control driver behaviour, tried to introduce speed limits, pedestrian zones, public transport and so on; but few of them linked their approach to air pollution reduction, and to air pollution network, with the exception of some cities like Strasbourg, Orléans and, to a lesser extent, Paris. Obviously, local governments were mainly pushing in the same direction as the transportation administration. Then from the beginning of the 1990s vehicle pollution becomes more and more central.

A Modification of the Policy Programme: Towards a Unification of the Two Arenas?

Several modifications of the clean air policy programme had occurred by the end of the 1980s, mainly under pressure from the EEC, but also, to a lesser extent, under environmentalist and social pressure. These changes are explained and reflected by a change in the composition of the policy network.

The first change was introduced by the EC directive which set an 'immission' limit for lead. This requirement imposed an adaptation of the monitoring system. There was a problem in financing the requested monitoring stations, considering that no representatives of car polluters as such were included in the financial organization of the networks. Central government decided to finance them, but to leave their management to local associations. The modification of the monitoring network can be analysed as a first attempt to introduce car pollution concerns into the air policy network, and it was the result of an incentive from the EEC.

The second change in the policy programme came with Decree (25/10/91) of 1991 which was enacted in order to integrate the EEC directives related to NOx, SO_2, lead and dust, into French law. The decree constituted another step toward the opening of the air policy network to car pollution problems in the sense that it enlarged the basis to be taken into account for defining special protection zones and alert zones. Such an enlargement is explicitly linked to vehicle emissions. In practice, this new policy programme has been recently implemented in the Ile de France region, with the modification of the content of the Protection zone (Arrêté 22/1/97). The new act explicitly specifies special measures for vehicle air pollution. The recent clean air law of 1996 constitutes the latest enlargement of the air policy programme. The changes introduced by the law seem to be more important than the previous ones, because they tend not only to extend the air policy network to car pollution, but also to enlarge the transportation policy network so as to include air pollution (by opening the PDU to air pollution, for instance).

The formulation phase of the new law has involved many more actors than the 1961 one. An *ad hoc* working group has been set up, composed of elected bodies, scientists, NGOs, representatives of industries and public transport, its aim being to contribute to the contents of the bill drafts and to advise on the different drafts produced by the ministry. The working group met half a dozen times and produced several drafts. One can assume that such an advisory group helped to change the perception of the problem by the different actors involved. The law states that air quality objectives, alert thresholds and air quality standards will be defined by decree in accordance with EU directives or in accordance with WHO recommendations. In order to comply with these standards, the new law imposes several important measures to be taken at local and regional levels.

The first measure deals with air pollution monitoring. The law imposes an obligation to monitor air quality in those urban areas with few industrial sources of emission (immediately for urban areas with more than 250 000 inhabitants; by the end of 1997 for urban areas with more than 100 000 inhabitants; and by 2000 all over the country). When the law was enacted, four cities out of twenty-one with more than 250 000 inhabitants were not equipped with such a monitoring system, and twenty-one cities out of forty-one populated with more than 100 000 inhabitants were not equipped. As a consequence, the new law will impose a significant financial burden to equip the region; 340 devices were considered necessary for such a purpose. The expected result of the monitoring effort is to strengthen the air pollution awareness of the populations of small and medium cities, in the same way as occurred in larger ones. A better level of sensitivity to the issue is also expected from local elected bodies. The latter could accelerate the rearrangement of the air policy actor network at local level.

The second measure deals with the implementation of air pollution reduction plans. Two different plans must be devised:

1. A regional plan for air quality: to be implemented for five years by the State representative at Regional level (Préfet de Région) which will define the main objectives to be fulfilled in terms of air quality objectives and in terms of emission reduction. This plan will begin with an inventory of emission sources, then it will define means to prevent and reduce air pollution. Lastly the plan will define specific measures for sensitive areas. The elaboration of the plan will involve the State administration at Regional level, the Regional government, and several regional and departmental commitees, composed of representatives of local governments, environmental NGO's, and qualified bodies. The population will be consulted through a public enquiry process.
2. A local plan for protecting air quality: it must be implemented in each city of more than 250 000 inhabitants and in those areas where air quality standards are about to be exceeded. This plan must also be elaborated by the State representative but at Departemental level (Préfet). The drafting process will follow the same consultative procedure as the one for regional air quality described above, but will involve the Departmental Government and representatives of different interests at Departmental level. This plan should be completed before July 1998 (according to the law). It will set up the main means to be used in order to meet air quality standards. These means are temporary or permanent restrictions to air pollution emissions from fixed or mobile sources. Special measures will be set up for emergency cases when alert thresholds are about to be exceeded.

At implementation level, these two categories of plans will extend the air policy network to car pollution problems. Both of them, try to deal with air pollution emission stemming from fixed and mobile sources of pollution in the same way. The realization of the plan will therefore inevitably involve actors with competence in car traffic (local mayors and DDEs) as well as actors with competence in air quality monitoring (monitoring associations) and air pollution management (DRIREs).

But the most important change introduced by the new Law is the reform of the PDU, which formerly was strictly a transportation policy tool (see above). The air quality law enlarges the objectives of these PDUs to include air pollution reduction and makes them compulsory for all urban areas with more than 100000 inhabitants, before 1 January 1999. The new PDUs are intended to reduce automobile traffic, and to increase public transport and less polluting modes of transport. But, contrary to the two previous plans, these PDUs are the responsibility of local governments (except for the Paris Region), in charge of urban transportation, with the collaboration of central government representatives and ultimately in collaboration with transport users' representatives and environmental NGO's. PDUs will again be submitted to public enquiry.

We can assume that such a reform will probably enlarge the transportation policy network at implementation level. Because they must be elaborated during the same period, one can assume that the new PDUs and the new air reduction plans process will require bringing together the air pollution actors and the transportation actors. However, many decrees must be enacted in order to implement the law. In November 1997, only three of them were about to be enacted; the agenda of the implementation might be very difficult to adhere to. None the less, the analysis of the enlargement of the air policy network at implementation level will be very interesting to study.

CONCLUSION

This chapter has shown how a policy problem perception and then a policy network have changed over time. The inter-relationships between both phenomena are obvious and clearly illustrated by the case of clean air policy.

However, whereas the past history is characterized by internal changes of perception, that is, the changes which are directly linked to the modification of actors position within the policy network, the more recent history can be characterized by an external change linked to the intervention of an external actor: the EEC. As a matter of fact, as shown in the chapter, the change of problem perception among air policy as well as transport policy networks, has been initiated by EEC directives. As a consequence, this supranational intervention has modified the configuration of the air pollution policy network in France, or at least has helped to modify it.

Further, in our investigation of situations both historical and present we realized that policy change occurs slowly: in the first case, the presence of two incompatible groups slowed down the process of change; in the second case, because of the fragmentation between the two arenas involved traditionally – those of the clean air policy and the transport policy one – the modification of the network at national level took a long period of time. And the modification at local level will be much more difficult.

The change of policy programme effected by the 1996 clean air law gives an opportunity to modify these local networks. Such an enlargement will be facilitated by the social and electoral pressure in those dense urban areas. As a matter of fact, the urban population seems to be more and more preoccupied with air pollution problems. Such awareness, which developed recently, is partly a consequence of better public information (here again induced by the EEC directives, especially the Ozone one).

However, as regards the effectiveness of such a policy programme, the legacy produced by twenty years without measures directed at car air pollution emission has serious implications. The changes of practices and drivers' behaviour will probably take as long as the change of problem perception took at the central level.

NOTES

* The introduction and the first part of this chapter were written by C.A. Vlassopoulou, while the second part and the conclusion were written by C. Larrue. A more detailed development of some of the arguments presented here is to be found in C.A. Vlassopoulou, 'Clean air policy in France and Greece', PhD in progress, University Panthéon-Assas, Paris II.
1. Similarly, the imposition of a new definition would reflect a transformation of the networks' configuration. French clean air policy constitutes a compelling case for the application of the definitional approach. During the past 40 years air pollution has passed through different definitions and networks. The decision to start our analysis at the beginning of the 1960s is linked to the fact that the first definition of the air pollution problem was formulated in 1961. The following pages concentrate on clean air policy at national level. In analysing the problem definition process during the last 50 years, we realized that the change in the definition of the clean air problem and the controversy in the public network it produces, are accompanied by the development of new laws which seem to reflect this restructuring. Thus we will first study how the clean air problem was defined through the first Clean Air Law of 1961 and the restructuring of previous policy networks.
2. See the French administrative directory 1970–74.
3. The term 'immission' refers to the quality of ambient air (it has a German origin).
4. For a more detailed presentation on the implementation of the French air quality policy, see Knoepfel and Larrue (1985).
5. The figures given here are from IFEN, (1995).
6. The preparation of emission standards has been heavily influenced by the problems of acid rain and of global climate change. This is not taken into account here because our chapter focuses on local and regional air policy. For an analysis of such problem transformation, see Roqueplo (1988).

7. Lyon's urban transportation policy and the air quality problem: a policy network approach

Grégoire Marlot and Anthony Perl

INTRODUCTION

This chapter deals with institutional resistance to the implementation of urban traffic control policies. The transport sector is today responsible for most of the urban air pollution in Lyon: 15 per cent of the SO_2 emissions, 28 per cent of the CO_2 emissions, 75 per cent emissions of total suspended particles (from diesel engines), 80 per cent of VOCs, 88 per cent NOx emissions, 95 per cent of the CO and 100 per cent of lead emissions (DDE, 1993).

Lyons' air pollution problems are exacerbated by two different transport patterns which coincide in the urban centre. On the one hand, the urban area generates its own traffic, both of passenger and freight transport. On the other hand, Lyon is at a nexus in the French road network, especially for freight transport and seasonal vacation travels.

There are no local policies against air pollution. The French state specified the maximum levels of emissions for each type of pollutant, and the Préfet (who represents the state's executive power equivalent to county level) had the authority to stop the industries if the maximum pollution levels were exceeded, but he hardly ever did it. Traffic reduction policies have recently appeared on Lyon's transport policy agenda as a means to combat growing air pollution problems, insofar as city officials do not have any other legal ways to control pollutant emissions.

Creation of an air pollution measurement system in the early 1990s established an analytical framework within which Lyon's air pollution problem could be put in perspective. Indeed, pollution crises are recurrent in big French cities such as Paris or Lyon, both in December–January and in July–August because of the specific meteorological conditions. It reached so high a level during recent years that it began to focus public attention on air quality issues. Television news and newspapers stressed the effects of air pollution, especially on the health of children and elderly people, publicizing the increase in respiratory complaints such as asthma and allergies. A national law on air pollution passed on 30 December 1996, provided an opportunity to restrict car use during air pollution crises.

TRANSPORT POLICIES IN LYON: A HISTORICAL PERSPECTIVE

Lyon has exhibited incredible political stability within the French context. Only five men have served as mayors since 1905. Such longevity enhanced the autonomy of local government (Benoît et al., 1994). Two mayors were particularly influential in shaping Lyon's transportation system: Edouard Herriot, who served from 1905 to 1957 (but not during World War II), and Louis Pradel, who served from 1957 to 1975.

From the Late 1940s to 1957

Despite Lyon's population density, city officials preferred to develop roads rather than public transport in the postwar years. Transport policy was premised on the assumption that cars would become the primary mode of urban transport in the future, a phenomenon that Yago (1984) has documented in Germany and the United States.

Traffic congestion appeared by the 1930s, and in 1935, the French state, the Rhône county and the city of Lyons joined forces to finance an infrastructure development programme. Two major roads were planned: an East–West throughway, with two bridges and a tunnel, and a North–South throughway on the right bank of the Rhône river, linked to the left bank by two bridges. Because of the war, the building of these facilities was not completed until the late 1950s. The city also financed the building of a modern ring road at the city's periphery, which had only a few connections with the urban road network and was over engineered.

When Edouard Herriot came back to head Lyon in 1945, he did not change any policies. From 1945 to 1957, when Herriot died and was replaced by Louis Pradel, the entire urban policy was a direct inheritance of choices made in the 1920s and the 1930s. During the 1950s, the mayor was mainly supported by Lyon's small business owners (Lojkine, 1974). Lyon lacked a strategic vision and long-range urban plan, since such political support oriented urban policy towards small-scale issues. In the transport sector, this lack of vision translated into a *status quo* approach to both road and public transport planning, whereas there was strong demographic and economic growth.

In the public transport sector services were managed by a private company, OTL (Office des transports Lyonnais). As a profit-making company, OTL always refused to build unprofitable lines, that is to extend lines into the suburbs. The Herriot administration sued many times during the 1920s to force OTL to expand services on routes with high demand but marginal profitability. Following these trials, the municipality financed the building and operation of unprofitable lines, while the others remained financed by the OTL. When the OTL franchise ended in 1941, the town council and the county council founded an association, the STCRL (*Syndicat des Transports en Commun de la Région Lyonnaise*), to take over OTL's responsibilities. Under the 1941 arrangements, the STCRL financed the whole public transport sector, and paid the OTL to manage public transport network operations until 1976.

Despite the introduction of this public agency into public transit financing, the elected officials who ran the STCRL were more concerned with balancing their budget than with building unprofitable lines in the suburbs. Neither the public officials nor their political supporters viewed public transport as a modern means of urban transportation, so that the STCRL was very reluctant to make new investments. Consequently, services did not grow to meet travel demand in Lyon's outskirts.

In 1956, the STCRL permanently abandoned trams and replaced them with trolleybuses. Lyon's elected representatives considered tramway operation too expensive; they thought indeed that people would progressively give up public transport in favour of the car, and so they wanted to avoid what they perceived as sunk costs. Following this logic of short-term profitability, public transport was deprived of exclusive infrastructure, so that buses and trolleybuses had to share the same streets as car traffic, and thus experienced the same risk of delays due to congestion.

Lyon's policy of disinvesting in transit was rapidly reflected in the evolution of modal choice. From 1945 to 1960, the annual number of journeys by public transport fell from 222 million to 168 million. At the same time, the traffic on the ring road rose 15 per cent per year during the mid-1950s, and doubled between 1959 and 1964. The length of commuting trips also grew. The number of people working in the city but living outside grew 50 per cent between 1951 and 1954.

Mayor Pradel's 'All-Car' Policy (1957–1975)

The political alliance between the mayor, the city council, and the county council permitted the financing of road development through the mid-1950s. The policy style of the new mayor, Louis Pradel, was very different from his predecessor's insofar as he had a long-term vision for the city. He wanted to provide new infrastructure and facilities in order to generate demographic and economic growth, since his aim was to make Lyon into France's second capital after Paris.

Pradel's ambitious vision identified mass motorization as an engine of urban development and prosperity. Automotive infrastructure would thus reinforce the 'central functions' of the city (tertiary activities and trade). Residential and industrial functions were assigned to the outskirts: the first in rapidly sprawling suburbs, and the second in warehousing and manufacturing zones. Pradel's transport plan was designed to permit easy commuting throughout the whole urban area by means of radial roads and ring roads, and to provide direct access to the urban core for intercity traffic by means of the North–South expressway.

Pradel's centrally focused road infrastructure plan did not please the mayors of surrounding towns, which would be reduced to the role of dormitory communities while they wanted to preserve their rural status. It was impossible to reach a consensus on Pradel's vision, because the mayors of small towns constituted the majority of the county council. The pattern of consensual infrastructure financing was thus broken, and along with it the initial postwar policy network pattern. This rivalry between Lyons and its suburbs resulted in the creation by the French parliament of the Lyon urban community in 1966, which was called *COmmunauté URbaine de Lyons* (COURLY). Since a majority of the COURLY representatives were elected from the city and inner suburbs, Louis Pradel gained a working majority for his urban development agenda, and a means of funding it.

Between 1958 and 1963, there was no abrupt break with the policy of Mayor Herriot. The completion of the building developed through the first postwar policy network cost FF40 million (mainly for the North–South expressway), only FF5 million of which was financed by the French state. To complement these investments, the city built 15 000 parking spaces in the city centre. Nevertheless, the congestion problem remained acute because no thought was given to diverting intercity traffic from the city centre, where it combined with local traffic. Small business owners – who constituted the main political support for the Pradel administration – were convinced that intercity traffic and its accompanying congestion was the price to pay for the economic growth of the city and their own prosperity.

From 1965, the creation of the COURLY provided financial means to implement the reengineering of the city centre and the development of 'new centres'. Lyon began to build a highway network of radial roads and ringroads, devoting a quarter of its budget to road development. The latter totalled FF125 million, even though the French state and the county financed most of the Fourvière tunnel (FF135 million out of 153 million). By comparison, the public transport budget was only FF22 million. The financial help of the French state and the county grew between 1970 and 1975: the first financed 50 per cent of the Perrache interchange, and the second, 25 per cent. The state also paid for the main part of the roads around the new centre of La Part-Dieu (FF48.7 million out of 52 million) and FF15 million out of 22 million spent on other projects.

Public transport was neglected during the 1960s as well as in the 1950s. In 1965, the public transport company TCL served only 150 km^2 in a 2 000 km^2 urban area. High-density suburbs with public housing units were very badly served by public transport – too few bus stops, low service frequency and slow trips. Low-density suburbs were not served at all. More than 30 per cent of the housing in the urban area was not served. Public transport thus lost many users (168 million journeys in 1960 reduced to 154 million in 1973). As a result, the STCRL began to lose money in 1963: FF17.5 million in 1972, and FF24 million in 1973 (Bonnet, 1975).

At the end of the 1960s, road congestion finally led city officials to increase investment in public transport. The TCL bought up two public transport companies that served the suburbs in order to expand the service; the underground railway construction – decided in 1963 – was launched in 1973. The oil shock does not seem to have played an important role in the decision to begin the works. In fact, the purpose of the project was not to provide a better public transport service, but rather to guide urban development (Lojkine, 1974) and to improve the city's image.

During 1968–1975, Lyons' transportation investments clearly favoured the car in spite of the huge cost of building the subway (see Table 7.1). Subway construction accounted for the bulk of public transport investments, along with the Perrache intermodal transfer station and shopping centre. The two projects show how Lyon's transport policy was not governed by welfare purposes, but corresponded to the political agenda of the mayor, which was to control the development of the city and to leave his stamp on the urban structure. International grandeur and prestige were the driving forces behind Pradel's transport planning (Lojkine, 1974; Bonnet, 1975; Benoît et al., 1994).

During 1975, Lyon's transport development was guided by a disjointed policy network, where local officials always opposed national state interests. Local public decision-makers had a high degree of autonomy in this policy network. They had technical capacity, with their own public transport experts, and benefited from the technical competence of the DDE (Direction Départementale de l'Equipement, which is a branch of the Transport Ministry) for major road projects without being compelled to follow its advice. Moreover, they were freed from political constraints, since decisions were highly centralized thanks to the COURLY's political structure. Although the French state financed a large part of the transportation investments, the Préfet and the DDE were never able to control the technical choices made by the COURLY. For example, the DDE opposed the building of the Fourvière tunnel and the Perrache interchange, but the Préfet did not use his power to enforce the DDE's recommendation.

Table 7.1 Transportation investments in Lyon, 1968–75

		FF millions
Private transport 60%	Urban highways	596.8
	Urban roads	628.7
	Perrache interchange (+ associated parking spaces)	30.0
	Parking spaces	70.4
	Traffic planning	22.0
	Total investments in private transport	1 347.9
Public transport 40%	Subway studies	22.1
	Subway works	663.5
	Funicular railway renovation	20.6
	Buses and trolleybuses	130.7
	Perrache interchange (building, shopping centre, access roads)	34.2
	Perrache interchange (bus and subway stations, taxi station)	42.7
	Total investment in public transport	913.7
	Total transportation investments	**2 261.6**

Source: (DDE, 1975)

Transport Policy Since 1975: Twenty Years of Ambiguity

The widening urban sprawl during the 1970s caused a growing imbalance between suburban housing, tertiary jobs in the city centre and industrial jobs scattered throughout the suburbs. The development of the road network, along with the inadequate public transport, encouraged travellers to switch from public transport to cars, all the more so since more people were able to afford a car because of the growth in their incomes. As a result, traffic jams dramatically developed.

The purposes of Lyon's transportation policy since 1975 appear very ambiguous. The urban development scheme (Schéma Directeur d'Aménagement Urbain or SDAU), which embodied Pradel's view of a city surrounded and crossed by highways, was never called into question. At the same time, four subway lines were opened between 1978 and 1993, and two others are under construction. Even though spending on traditional public transport (buses and trolleybuses) only represented 23 per cent to 30 per cent of the total investment in public transport, the service was improved and the network developed (28 new lines since 1978).

According to TCL, the annual number of trips rose from 146 millions in 1977 to 216 millions in 1993, of which 54 per cent were by subway. Nevertheless, the improvement of the public transport network failed to halt the growth of car use (Marlot, 1996). The subway gained users in the city centre, while travel by bicycle, motorcycle and on foot declined at the same time. The suburbs are served by buses (there are only three subway stations in the suburbs) which suffer from the traffic congestion. Public transport service remains very slow and insufficient in these areas. 28 per cent of the households were not served in 1972, dropping to 16 per cent in 1987 (SYTRAL 1986). As a result, surface public transport never stopped losing users even after its improvement. Tables 7.2 and 7.3 illustrate the trends for urban mobility in Lyon since 1975.

Table 7.2   Mobility evolution by transportation mode: Lyon, 1976–96

	1976		1985		1996	
	Trips*	%	Trips	%	Trips	%
All transportation modes	3.75	100	3.48	100	3.9	100
Walking	1.71	45.6	1.2	34.5	1.22	31.3
Bicycle	0.08	2.1	0.04	1.1	0.03	0.7
Motorcycle	0.11	2.9	0.03	0.8	0.02	0.5
Public transport	0.41	10.9	0.49	14.0	0.5	12.8
Private car	1.42	37.8	1.70	48.8	2.10	53.8
(of which lone driver)	*1.07*	*28.5*	*1.32*	*37.9*	*1.62*	*41.5*

* Number of trips/people/day.

Based on data from the (Ministère de l'Equipement et des Transports (1976, 1985, 1996)

Table 7.3   Evolution of commuting trips within the metropolitan area of Lyon, 1976–82 (trips per day)

Location	1976	1982	%
Within Lyon/Villeurbanne	215 000	175 000	–19
Within the SDAU (excluding Lyon/Villeurbanne)	128 000	138 000	+8
Between Lyon/Villeurbanne and the rest of the SDAU	124 000	138 000	+11
Total	467 000	451 000	–3

Based on data from the Ministère de l'Equipment et des Transports (1976, 1985)

THE POLLUTION AND CONGESTION PROBLEM: A POLICY NETWORK APPROACH

Structures of the Transport Policy Network

Public and private transport policies are planned, implemented and financed by separate groupings of state and societal actors. This disconnection between the policy networks has frustrated all solutions to date, since problems requiring coordination between them, such as traffic reduction or air pollution, prove exceptionally difficult to resolve. The automotive policy network is well advanced in the implementation phase of the SDAU which was an answer to the need for urban planning expressed by the French state in the late 1960s. It was debated for nine years by the French state, the county council and the COURLY, and was finally accepted in 1978, largely unchanged from its 1969 version. Mayor Pradel's transportation policy followed a long-term vision, which was taken over by the SDAU and continued to the present. It relied on dated technical choices and inflexible institutional arrangements, and has proved very difficult to adapt to changing priorities, such as the rise of environmental concerns. By comparison, Lyon's public transport policy network is more flexible, although also lacking in long-term vision. Subway lines were built without an overall development scheme; for example, the SYTRAL (the former STCRL) recently decided to extend a subway line into a declining industrial area in order to serve the Soccer World Cup stadium. Moreover, the relationships between the participants of the two policy networks are different. In the public transport network, SYTRAL, which is responsible to the city council, takes decisions on planning and finance, while other actors are only service providers. By contrast, the highway transport policy network requires considerable negotiation to reach an agreement, because the interests of the actors often diverge.

Transit and highway policy networks are also divided by separate financing arrangements. The national government finances 50 per cent of major projects, such as the eastern bypass road, and the COURLY and the county each pay 25 per cent of the costs. Fiscal limitations have recently led public officials to franchise the development and operation of new highways to private companies. By contrast, the COURLY finances only public transport, and its financial constraints dramatically limit the opportunities for major development projects.

The public transportation policy network

The institutional landscape of public transport suggests the complexity of the relationships between the actors who plan the transportation, who ratify policy choices and who implement them. The policy network involves experts both from national and local state organizations, public authorities and private companies.

Users are only indirectly represented in the policy network by elected representatives of the COURLY and the county council.

Overlapping roles and responsibilities have clouded decision-making in the public transport policy network. The SLTC (Société Lyonnaise de Transports en Commun), a private company which manages the public transport network, is a subsidiary of Transexel, a corporation managing public transport in several major French cities. The SEMALY, Société d'Etudes du Métropolitan de l'Agglomeration Lyonnaise, which is chaired by the mayor of Lyon, is a mixed enterprise which builds the subway. The AGURCO (AGence d'URbanisme de la COURLY) is a local bureaucracy producing transport studies. The DDE is a branch of the Transport Ministry and the CETE (Centre d'Etudes Techniques de l'Equipement) is the DDE's research department. The councillors of the city, the county and, since 1983, the region, are the elected representatives who take the decisions.

Since 1983, the SYTRAL (SYndicat des TRAnsports Lyonnais) has replaced the STCRL. As well as the AGURCO, it is led by the COURLY's representatives as required by France's 1983 decentralization laws. Its functions are to plan and implement the public transport policy; it is an expert and a decision-maker at the same time. The SYTRAL recruited its own technical staff in order to counter the TCL and SEMALY agencies' expertise (Offner, 1990) and, as a result, developed into an autonomous local bureaucracy with technical and financial capacities. Because the decision-making power rests on strategic data, which are mainly produced and used by the bureaucracy who implement the policy (Niskanen, 1971), the SYTRAL is the key actor of the network. SYTRAL's engineers do not depend on the data provided by the TCL and the SEMALY. They are unelected and relatively insulated from public pressure. SYTRAL is under the supervision of the elected representatives of the COURLY and the county council, who approve the public transport budget but lack the ability to critically appraise transit policy. SYTRAL's engineers also have an intermediary function between the elected representatives and the TCL and the SEMALY, who now negotiate their operating contracts through the SYTRAL.

Figure 7.1 illustrates the relationships between the companies who operate public transport services, the different elected representatives and the bureaucracies at different levels of Lyon's public transport policy network. All these actors are interdependent: the TCL and the SEMALY have only one client and depend on the COURLY's money; but the COURLY also needs these firms to implement its public transport policy. The COURLY decides the SYTRAL's budget, and the SYTRAL must work with the TCL, because of its longstanding franchise, as well as with the SEMALY, which is linked to the COURLY. Even if the SYTRAL holds most of the decision-making power, it relies on the other actors and is obliged to negotiate with each of them.

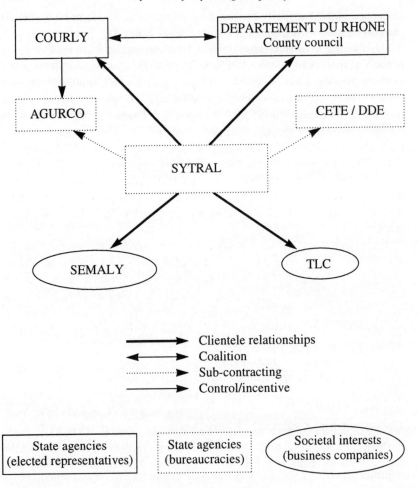

Figure 7.1   Lyon's public transport policy network

The policy network perspective shown in figure 7.1 suggests that the public transport policy network faces a problem of competition among state agents at different levels which compounds the difficulty of forging effective relationships between societal and state actors. Amid the multiplicity of state actors, SYTRAL's coordinating role gives its leaders political autonomy. As a result, the SYTRAL's managers implement a public transport policy consistent with their engineering culture, which values increasingly technically advanced services. A good example is the subway line 'D' which was needlessly equipped with an automatic driving device that cost millions of francs. SYTRAL thus appears as an autonomous organization that fills the vacuum created by the dispersion of state authority and the low organizational development of societal interests. State officials, both at local level (COURLY, county council, AGURCO) and at national level (DDE, CETE), are unable to act independently of the SYTRAL. They require its data and expertise in order to coordinate policy network activity. Thus SYTRAL acquires an opportunity to promote its own point of view. By Skogstad and Coleman's (1990) definition, Lyon's public transport policy network appears as a 'clientele pluralist' policy network, since SYTRAL manages a diverse set of public and private 'clients'.

The highway transportation policy network
The highway transportation policy network comprises strongly organized societal interests who advocate policy priorities to a few state agencies with differing interests. This combination of divided state authority and powerful societal interests groups gives rise to a fluid mode of state–society interactions typical of pressure pluralist networks (Coleman and Skogstad, 1990a). Groups approach state agencies and actors issue by issue, lobbying for specific interests. The resulting interactions are enlivened by the competition among state actors. Some of these state agencies seek to channel social pressure to promote their own interests, while an agency such as the DDE tries to protect the national government's interests.

The key decision-makers are the COURLY, for the building of urban roads and ring roads, and the DDE, for major national roads. Their relationship is discordant. The DDE, which also provides its technical abilities and advice to the COURLY, favours national interests such as a unified national road transportation scheme. By contrast, the COURLY favours local interests, that is infrastructure which generates urban economic growth. The interest of the DDE and the regional council are generally convergent, as are the COURLY's with those of the county council. Unlike public transport, highway projects capture public attention, usually generating negative political input. People organize themselves and pressure local elected representatives to stop highway infrastructure in their own neighbourhoods. Consequently, infrastructure development generates political bargaining between technical agencies and local elected representatives and the latter act as a channel for the societal interests to pressure state actors. Nevertheless, the societal pressure on decision-makers is never very high, since ideology and party solidarity serve to insulate suburban mayors and other public officials from public protest.

Organized city centre citizens pressure urban public officials to address noise and congestion problems; even so, they are not ready to abandon their cars even if a large number of car trips in Lyon are shorter than one kilometre. They are supplemented by a silent majority who rarely take note of transportation issues, but want above all to keep the ability to use the cars. As a consequence, most residents favour building new roads outside the centre, such as ring roads

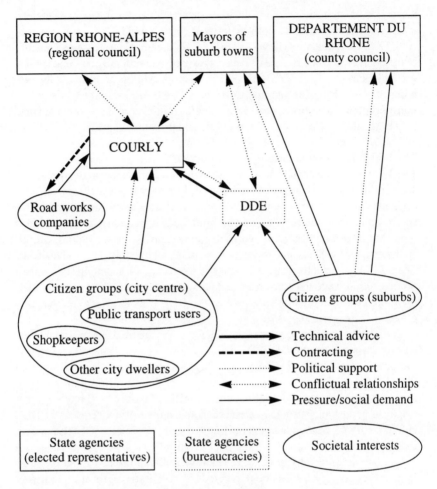

Figure 7.2 Lyon's private transport policy network

and bypass roads. Public transport users advocate the extension of the subway, especially to their own district (the bus is deprecated). City centre traders oppose traffic regulation measures because they fear losing customers and advocate building new parking spaces, although some of them now favour pedestrian areas because these are seen to increase business.

Engineers and planners in government place top priority on improving the flow of through traffic. Smooth traffic flow is seen as a matter of prestige, because for years Lyon has been famous for its traffic jams, especially in the *Fourvière* tunnel. This is an economic priority to enhance logistics. Expediting through traffic is also a political priority, because many city dwellers are fed up with the traffic jams caused by through lorry traffic. Moreover, the lorries make a large contribution to air pollution problems (54 per cent of NOx emissions; DDE, 1993). Consequently, the COURLY has concentrated on building bypass highways: the eastern and the southern bypass roads were opened in the early 1990s, approximately 20 km away from the city; the western bypass road is still on the drawing board, since it faces great public opposition.

Suburban societal input has significantly shaped transportation policy over the last ten years. Like suburbanites around the world, Lyon's *banlieusards* live where they do to avoid the high prices, noise and pollution of the city. When the COURLY and the county council propose a new bypass highway, which crosses the suburbs and generates the very noise and air pollution that suburbanites sought to escape from, we find a typical 'Not In My BackYard' (NIMBY) effect. Citizens mobilize locally to pressure their local elected representatives to reject the proposed infrastructure. Local representatives are then caught between the demands of their constituents and the advantages of the new infrastructure, since the reduction of travelling time can raise property values by increasing residential in-migration as well as corporate relocation.

Suburban public officials face the challenge of gaining benefits from building new road infrastructure without risking retribution from their constituents at the next election. In practical terms, they negotiate with the COURLY and the DDE as if they do not want the infrastructure (for example an interchange or a highway exit on the edge of town), in order to get compensation. From their electors' point of view, the suburban politicians have been forced to agree to the building of road infrastructure, but have succeeded in gaining compensation. Local officials are thus not seen as responsible for the noise and pollution, even though local government will reap the benefit from the road in terms of more residents, more companies, and thus more taxes.

The bypass roads that have been built around Lyon gained many suburban interchanges because of such deals between the COURLY and the suburban mayors. Consequently, the bypass roads that were intended to divert through traffic from the city now serve as access roads, stimulating urban sprawl and traffic. For example, the eastern bypass is now congested only three years after its opening because its many entrances and exits stimulated commuter travel. This outcome fits the global trend in expressway expansion in and around cities (Newman and Kenworthy, 1988).

Institutional and Technical Constraints in the Field of the Transportation Policy

Institutional constraints on traffic reduction measures

Lyon's disjointed urban transportation policy networks have created both processes and outcomes that are quite unpromising for air quality management initiatives. At a procedural level, there is no motivation among local political and economic elites to define urban transportation as a major cause of Lyon's air quality problem. Nor is the general public mobilized to demand such a change in the public agenda.

Despite a lack of public opinion surveys about air pollution, public support for air pollution control policies appears weak. Societal groups are far more likely to mobilize against the noise and the visual externalities created by new infrastructure, or to protest about traffic congestion, than to advocate air pollution control policies or traffic reduction measures. Public information about air quality problems (provided by the COURLY) is conservative. Pollutant levels are always reported below maximum levels which are not defined. Consequently local newspapers play down these figures, and local television reports air pollution measurements, when audiences are small. By contrast, the DDE and the local officials receive many complaints concerning noise from road infrastructure; Lyon's business interests regularly demand improvements in traffic circulation; and congestion in itself seems to represent a demand for more roads.

The traffic reduction measures that would help clear Lyon's air are opposed by shopkeepers, who fear a reduction of their clientele, and by corporations, who fear making their employees' journeys to work more difficult. Because Lyon's outer suburbs are almost entirely 'auto-dependent' (Newman and Kenworthy, 1989), their residents and elected officials both oppose restrictions on vehicle use, since this would radically reduce property values at the same time as it increased the difficulty of accessing jobs in the city centre. State actors like the DDE do not consider traffic reduction measures to be their responsibility. On the contrary, their engineering training leads civil servants to propose technical solutions to both transportation and environmental problems. Regulatory or pricing options remain off the policy agenda, even if the DDE is the only state agency which pays attention to air pollution, as well as noise problems, resources consumption or traffic congestion.

Technological constraint on transportation policy

Lyon's transportation policy has created technological constraints on adaptation, insofar as the car and public transport are both network technologies. These network technologies display increasing returns to adoption: the more they are adopted, the more efficient they become. Unlike the allocation process in a standard economic approach, the selection process between two technologies with increasing returns is unpredictable. Agents are unable to choose rationally between technologies, because they can not make good predictions about the evolution of the technology's returns. As a consequence, their choice is uncertain, since each stochastic decision in favour of one technology increases the probability that the next agent will choose that same technology. Thus, the particular sequencing of choice made close to the beginning of the process strongly influences the whole selection process, as the final outcome can be influenced by expectations built up early on.

Three types of interdependence structures generate positive feedbacks: dynamic returns to scale (for example, learning by using), positive network externalities, and interdependencies linked to the systemic nature of some technologies. The technical improvement of cars and public transport systems depends on the number of users. The more a transportation mode is adopted, the more experience people gain with it. Scale dynamic returns linked to learning by using effects appear at a global level for cars, but only at a local level for public transport. The automobile has confronted many problems (safety, congestion, parking, fuel consumption and harmful effects such as noise or air pollution) which have stimulated many technical improvements. Moreover, the small number of increasingly global car manufacturers can direct considerable resources into research and development and disseminate the results widely. By contrast, each city has its own urban public transport system. Each transport system is specific to the city and is managed by a local or at best a national firm. Therefore, buses, tramways and subways do not benefit from learning by using effects at a global level (there are no scale economies, no centralization of experience) and R&D efforts cannot be quickly and widely diffused.

Positive network externalities also appear at a global level for cars, but at a local level for public transport. Both technologies enjoy a public goodwill, either because their use requires the use of a related public good (that is, individuals' driving requires the use of a road network and parking space) or because they are operated as a public utility (buses, for example). The more users adopt a transport mode, the more that particular network develops, so that when an individual chooses one transport mode over another, he produces a positive externality for all other users of this mode. For example, increasing auto-mobility is a great incentive for public officials to invest public funds in expanding the road network. Conversely, when more people commute by public transport, it provides more money to operate the network, as well as a justification for city officials to invest public funds in expansion.

The systemic nature of transportation technologies creates positive feedback effects between the development of a transport network and specific forms of urban growth. The transportation mode which is gaining market share influences urban development. Cars are inefficient in very crowded urban areas and their use generates pressure for decentralization, since they are more efficient in low-density suburbs. By contrast, mass-transport systems create opportunities to move high volumes into urban areas, enabling high density living and working arrangements.

The need for air pollution control measures and traffic reduction policies

Despite the social and technical irreversibility of the automobile. Lyon's transportation policy is now facing a crisis of automobility. After 50 years of road development in and around the city, congestion remains unresolved. Furthermore, transport has become the main air polluter (see Table 7.4), passing industry which was far more polluting in the early 1960s. At last, infrastructure is more and more expensive, while public transport is less and less profitable, so that the transportation policy has to tackle serious financial problems.

Table 7.4   Air pollution in the COURLY, 1990 (tonnes)

Pollutant emissions	Industry	Housing	Transport
SO_2	4072	1277	2660
NOx	1656	978	37104
Particles	374	171	905
CO	895	4023	152584
CO_2	1135135	1254806	1408967
VOC	1118	457	21710
Lead	0	0	50

Source: DDE (1993)

These three crises (congestion, pollution and finance) could force change in transportation policy. The financial crisis led to the introduction of new actors in the road policy network. Since public authorities lacked the money to finance the building of the northern ringroad, they granted the infrastructure to a private consortium which established a toll to get its investment back and make a profit. It also demanded a profitability guarantee, that is a reduction of other roads' capacities to ensure people will use the tolled expressway. This new financial logic introduces economic criteria into road use and reduces the amenities for those who are unable to pay. Be that as it may, public opinion strongly expresses its opposition to such methods, so that the toll will be reduced by 50 per cent. Congestion and pollution crises also create opportunities for societal or state actors to raise problems like air pollution within transportation policy networks. The December 1996 law on air quality will introduce the Environment Ministry as a new actor in the transportation policy network, thus creating an opportunity to make road and public transport policy networks more consistent.

INSTITUTION BUILDING OPPORTUNITIES FOR LINKING TRANSPORTATION AND URBAN AIR QUALITY IN LYON

Existing transportation policy networks are unlikely to implement traffic reduction schemes. Policy actors are not motivated to pursue a linkage between transport and air quality because state actors see limited returns on such an investment of economic and political resources. At the same time, societal actors – including the general public – have not identified this issue as a priority.

Reshaping the legal framework that allocates policy responsibility for transportation and urban environmental matters offers a good opportunity to transcend existing limitations. The *Loi sur l'air et l'utilisation rationnelle de l'énergie* opens the door to such legal reform by identifying several measures linked to traffic control. But, for this linkage to take effect, a traffic regulation scheme that is both politically acceptable and technically efficient must be designed and implemented.

What an Effective Traffic Regulation Policy Needs to Achieve?

Using pricing to control road use is widely recognized to be politically difficult, mainly because it prices something previously perceived as free and introduces a segregation between the users. Nevertheless, economic theory assumes that enough revenue can be generated to more than offset the costs to individual travellers. Some economists suggest that it ought to be possible to design a package of congestion charges and revenue uses, that looks attractive to most people, and thus enables policy implementation (Small, 1992; Giuliano, 1992). For example, Small proposes distributing toll revenues to reimburse travellers directly, to substitute for general taxes that now pay for transportation infrastructure, and to develop new transportation services. Besides avoiding the regressive impact of tolls by providing alternatives to people who do not have money enough to pay the toll, Small's distribution scheme has the potential to buy public support for road pricing.

An alternative approach that avoids the equity dilemma of road pricing is a regulation system based on a property right mechanism (Coase, 1960; Baumol and Oates, 1988). Following Coase's approach, congestion can be ascribed to the absence of clearly defined property rights for the use of urban road infrastructure. A regulatory system which gives people a limited access to urban road infrastructure would clarify those rights equitably. To maintain fairness, access permits would be distributed to all car-owners of the conurbation. For example, each driver could be allowed to drive in the city centre four days out of five during the working week, as proposed by Perl and Han (1996). Such a system not only reduces congestion, air pollution and noise, but it also compels people to use alternative modes of transportation one day per week. It progressively generates new habits of commuting, thus increasing the number of regular users of public transport, leading to improved profitability and service quality.

If driving permits were made traceable, it would allow some people to pay for the opportunity to use their car whenever they wanted to. To operationalize this policy, local government could organize a permit exchange, where people could easily sell and buy rights at the market price, using a telematic system like Internet, the French ' Minitel', or just a phone call. Such trading increases incentives to switch from the car to public transport, insofar as people who sell their access rights get a financial surplus which permits them to pay for public transport. The system is inherently fair, because everybody has the same right of access. If nobody wants to sell his rights, the affluent will not have more rights than the poor. Moreover, the number of access rights for sale is an indicator of the evolution of mobility habits: if there are a lot of access rights for sale, it enables public authorities to buy back these rights. Through such an iterative process, it is possible to develop a systemic reduction in car use and a corresponding renaissance of public transport.

Institutional Building for Traffic Reduction Policies and the French 'Clean Air Act'

The *Loi sur l'air et l'utilisation rationnelle de l'énergie* stipulates that every city of more than 250 000 inhabitants must set up an air pollution measurement system (which already exists in Lyon). The Préfet must inform the population when the air pollution level is too high; furthermore, he has the right to restrict car use during the pollution crisis by up to 50 per cent, by ordering cars with odd registration numbers off the road on one day, and those with even numbers the next day. Public transport would be free during the pollution crisis when the use of cars is restricted. Following the Italian examples (Desideri and Lewanski, 1996), we can see how this kind of emergency context creates opportunities to focus public attention on the pollution problems caused by car use, and thus to make people more likely to accept traffic reduction measures. Moreover, the law stipulates that the city transportation plan (*Plan de déplacements urbains*) must pay attention to the problem of air pollution and must be designed in accordance with all other plans, especially to the regional air quality plan (which currently exists only in Paris). The limit of such a measure lies in the Préfet's will and ability to enforce the law and the possibility of reaching a consensus between the actors. In Lyon Louis Pradel built infrastructure during the 1970s, such as the *Fourvière* tunnel, that was not in accordance with the plans designed by the DDE without any intervention from the Préfet.

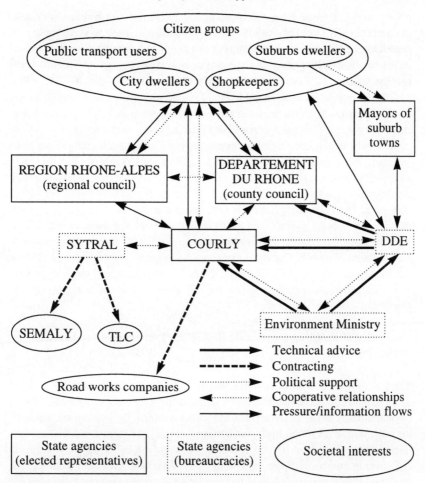

Figure 7.3 An integrated transport policy network

Integrated strategies of traffic reduction need specific institutional building to be implemented. Firstly, since traffic reduction implies a switch from cars to other transport modes, the city will have to develop transport facilities to meet new demands resulting from traffic reduction measures. Cooperation between the different decision-makers from both private and public transport policy networks (the SYTRAL for public transport; the COURLY, the county council, the regional council and the DDE, respectively for urban, local and national road infrastructure) will be needed to optimize an intermodal transportation scheme.

Secondly, national policy actors like the DDE and the Environment Ministry can facilitate compromise among local regional actors. Finally, the integrated policy network must be more open to societal input. As Swiss experience demonstrates (Schenkel, 1995), traffic reduction measures work best as a democratic process. Public awareness can be raised through several types of information campaigns: press coverage, advertising, organization of local forums to discuss the issue and explain the need for traffic regulation measures, and public debate between experts.

8. Clean air and transport policy in Switzerland: the case of Berne

Daniel Marek

INTRODUCTION

Questions and Observations

The idea of COST-CITAIR Action 618 is to improve the understanding how different institutions concerned by the control of air pollution work. The present case study of the city of Berne[1] examined the institutional arrangements of selected measures for air pollution control in the transport sector implemented between 1987 and 1994.

Because the implementation of air pollution control policy can hardly be understood as a hierarchically controlled process, this study applies the theory of social networks as a model and as an analytical tool. Implementation of air pollution control policy is seen as a result of the interaction of different administrative and political actors.

The emergence of policy interaction and informal cooperative structures between authorities and pressure groups (Marie and Mayntz, 1991; Kenis and Schneider, 1991) has resulted in increased attention being paid to the application of the theory of networks to the planning and implementation of policies. This approach can also be used to record the generally varying concepts of cooperation and competition between political actors (Jordan and Schubert, 1992). In the case of policy networks the theory works on the basic assumption of a decentralized coordination of the individual actors. This contradicts the formal constitutional viewpoint which assumes hierarchical control and the exclusive participation of administrative actors.

The examination of the empirical material led to surprising results. There is only a weak correlation between the structural characteristics of policy networks and the success of the concrete measures. Between the structure of policy networks and the measures ('output') exists a third factor: the negotiation process which is too complex for a simple causal model. But there are some soft factors which enable the success of the measures.

The empirical findings refute the existence of a negative link between the institutional establishment of the network (in the form of working groups or coordination committees) and the time necessary for the implementation of the measure. The participation of various committees in Berne led to a delay which became apparent in the policy development of the examined measures: the preparation of the measures tended to take several years. An important reason for this delay could be that the creation of coordination committees in the example of the city of Berne tends to be accompanied by an extension of the networks. The increased time requirement would appear to be the price to be paid for the institutionalization of policy networks which was necessary for the integration of highly specialized actors.

It is impossible to observe any connection between the degree of centrality and the extension of the measures, even in the case of behavioural measures such as zones with reduced speed limits or the introduction of a parking place management system (see below). More decentralized networks are not more successful in terms of implemented measures.

On the other hand, the empirical study substantiates the negative influence of the size of the network on the scope of the measure. This should be viewed in connection with the statement on institutionalization above. It was shown that the consultation of numerous actors hindered consensus. For example the inclusion of new actors led to a smaller 'perimeter' of some measures in Berne. This finding contradicts the findings of earlier studies in Switzerland (Knoepfel, Imhof and Zimmermann, 1995, pp. 119–21), who stated a positive link between the number of actors and the number of implemented measures. Generally there should exist an optimum size for networks; but this remains to be explained in theory. This optimum size should depend on the nature of the measure and the distribution of the resources among actors.

Research Strategy and Research Design

The institutional requirements for the success of measures for air pollution control in the city of Berne were examined on the basis of four concrete measures (called 'objects') which have actually been implemented in the transport sector. The selection of the measures was based on a assumed contribution to the improvement of air quality.[2] The entire case study is based on the examination of four measures or objects which were implemented in the city of Berne between 1987 and 1994:

1. the introduction of speed limit zones ('Tempo 30') in residential areas;
2. the provision of bicycle stands and shelters as an example of the promotion of alternative means of transport;

3. the distribution of residents' parking permits as a parking management measure;
4. a collective fare agreement for public transport in Berne and surrounding areas as an example of the promotion of public transport (called 'Bäre-Abi').

Five groups of variables have been examined with the aim of explaining the implementation of the selected measures: political programme; institutional characteristics of political and administrative actors; institutional characteristics of social actors; situative variables; and structural variables: (Structural variables are immutable framework conditions, for example tax revenue, employment structures, public transport amenities, pressure caused by stationary emissions and the residential structure of town centres and agglomerations.)

The relationships between these variables are organized in a dual-phase model: The first three groups of variables explain the structure of the policy networks. With regard to the structure of the networks this study concentrates on formal decision-making competencies, the availability of resources and the institutional power which results from the above-mentioned variables. Knowledge and reputation or legitimacy would also be viewed as resources of the individual actors. It would also make sense to survey other relationship classes (Pappi, 1987, p. 13). Due to the practical limitations imposed by the research process (surveying techniques, data quality), the selection made here is restricted to the most important classes.

The structure of the implementation network with its patterns of cooperation and conflict, actors' interests and positions of influence serves to explain the process of the realization of the measures or objects ('policy development'; see Schneider, 1988, p. 29). To facilitate a dynamic analysis no clear distinction was made between programming and implementation. The focus is more on the overall realization of an object than on the programming and implementation. However, proposals for solutions and perspectives of the problem which tend to be subordinated during 'policy development' must not be overlooked. Dynamic policy-output features are used here as a basis for the analysis of policy development. These features include the 'perimeter' or spatial scope of the measures, the scope of intervention (nature of the instrument which is intended to bring about a change in behaviour), and the scope of the problem (nature of the problems to be overcome with the help of the measures to be introduced).

It should be noted that it is unlikely that all of the variables mentioned here can be applied formally. Most of these variables concern qualitative features whose characteristics elude precise definition. The empirical research was based on a qualitative analysis of documents and a series of expert interviews as 'focused interviews' (Merton and Kendall, 1979, p. 179) with representatives of the involved actors.

THE CASE OF BERNE

General Conditions (Structural Variables)

As the centre of a medium-sized agglomeration with around 400000 inhabitants the city of Berne is subject to the typical problems arising from a different development between the town centre and the suburban area. The centre of Berne provides many jobs in the services sector; as a capital city it provides approximately 130000 jobs, of which 14000 are based in the cantonal and federal administration. The centre of the city alone is visited by an average 46000 people daily (working-day average). Like all cities in Switzerland, the city of Berne is now facing the consequences of the increasing spatial dissociation of living and working. Since 1970 the number of commuters from outside the city has clearly risen in relation to the numbers who both live and work within the city. In addition the city of Berne now acts as a centre for a wider region which extends to the towns of Biel, Solothurn, Burgdorf and Thun.

The increasing volume of traffic arising from the development of the city's economic structure, which is strongly rooted in the tertiary sector, represents one of the main sources of environmental pollution. Industry plays a subordinate role in the generation of air pollution, with the exception of the waste incineration plant and a few larger plants to the west of the city. Equally characteristic of the difficulties between town centre and agglomeration authorities is the relatively weak level of regional cooperation. The city of Berne is forced to cover the costs not only of the environmental pollution caused by commuter traffic but also of numerous other services and does not receive full compensation from the suburban authorities. Cooperation between the various political authorities in the region is loosely institutionalized in the form of the Association for Cooperation in the Berne Region (Verein für die Zusammenarbeit in der Region Bern, VZRB) which was founded in 1963. However the association is mainly involved in planning issues that affect the region.

Political System, Financial and Political Parties (Situative Variables)

Berne's finances tend to reflect general economic trends in Switzerland. Following more than a decade of budget surpluses, strongly increasing deficits began to emerge from 1990 onwards (Präsidialdirektion der Stadt Bern, 1992, p. 209). This gave rise to a sharp change in financial trends and the resulting effect on the city's businesses must be taken into account. During the period in which this study was undertaken the city parliament (*Stadtrat*) was characterized by widespread fragmentation. From 1985 to 1988 the combined votes of the left-wing and green groups and a few centralists could, theoretically at least, constitute an absolute majority. The fragmentation process continued, however, in the subsequent legislature of 1989 to 1992 when new groups, for example the 'Auto-Partei' (Automobile Party (AP), later Freiheitspartei) entered parliament at the extreme right of the political spectrum. The city government (*Gemeinderat*) was dominated during both legislatures by the bourgeois parties which had four of the seven seats. The Social Democratic Party (SP) had two seats and the remaining seat was initially taken by an unofficial Social Democratic candidate and later by a representative of the alternative-green groups.

In this context we should note, that in an international perspective the Swiss federal system displays a special distinctive feature. Generally in the field of environmental and transport policy the responsibility for the implementation of concrete measures lies with the cantons and the communities. The federal legislation sets only the general conditions for the actions of cantonal and especially local authorities. There are some similarities between the situation in the United States of America and Switzerland (which actually adopted several elements of the US constitution in the middle of the nineteenth century).

Air Pollution

Traffic is the main cause of current levels of air pollution in the city of Berne. There has been a recent reduction in the levels of nitrogen oxides (Fürsorge- und Gesundheitsdirektion der Stadt Bern, 1992; Präsidialdirektion der Stadt Bern, 1987–1992). However at 47 mg per m^3 (mean for 1993) emissions of nitrogen oxides were still in excess of the authorized ambient quality standards specified in the federal air pollution decree (Luftreinhalteverordnung, LRV); in the case of sulphur dioxide, which mainly originates from heating systems and industrial furnaces, a decrease in emissions has been recorded and levels have decreased from 15 mg per m^3 in 1989 to 10 mg per m^3. The efforts to control emissions from furnaces in recent years have succeeded. In addition to direct pollution caused by nitrogen oxides ozone levels are also critical; nitrogen oxides and volatile organic compounds are precursor pollutants in the ozone cycle. The levels of ozone continue to rise and in 1993 the authorized limit was exceeded 85 times in the average hourly levels.

Environmental and Transport Policy in the City of Berne

The current transport plan for the city of Berne was published in 1982 and 1983. This plan declared a multiple commitment to 'reducing the negative effects of today's traffic levels' (Gemeinderat der Stadt Bern, 1982, p.1). As part of this plan the city council aimed at reducing passenger vehicle traffic by restricting traffic access to a so-called basic network of roads. It was intended that these measures would stabilize and prevent any further increase in traffic volume (Gemeinderat der Stadt Bern, 1982, p.2). Shopping and tourist traffic were expressly omitted from these targets and the traffic restrictions applied to commuter traffic only. The shopping zone in the centre of Berne is very important for the city and tourism is also a major economic force. The conflict of interests arising between environmental concerns and the potential targets of the measures introduced can clearly be seen in this traffic plan. In the end, the implementation of the plan concentrated on relieving the burden of traffic on the residential areas through multiple efforts to improve and promote the public transport system and the management of parking amenities. An additional plan was drawn up for the management of parking resources, which mainly concentrated on the transformation of the approximately 17300 long-term parking spaces into short-term parking spaces or parking spaces reserved for residents. The idea behind this strategy was to deprive commuters of parking amenities and make them available instead to visitors and residents (Gemeinderat der Stadt Bern, 1983, p.27). From 1981 to 1987 twelve projects for widespread traffic control which aimed at reducing emissions from through traffic were implemented in residential areas. The projects mainly involved technical measures in street design as there was no national (federal) legal base for the introduction of speed limit zones (Tempo 30).

During the second half of the 1980s the question of air pollution was the focus of much attention in the context of *Waldsterben*, forest damage or dieback. Several public appeals were submitted to the city authorities condemning environmental pollution caused by traffic and demanding the relocation of traffic routes. In 1991 the canton of Berne published its air pollution measurement plan (KIGA, 1991) for areas with excessive levels of pollution, which is required by the Federal Air Pollution Decree (*Luftreinhalteverordnung*, LRV). The measurement plan contains procedures to reduce the volume of traffic in the city through the management of parking amenities and the promotion of public transport, and is thus mainly based on existing transport planning for the region of Berne. The plan also contains additional traffic control and economic measures.

The next step in the process involved the presentation by the Berne city council of an urban development concept (*Stadtentwicklungskonzept*, STEK). With 'economic eco-city' as its slogan this concept aims at the targeted environmentally-friendly development of the city (Gemeinderat der Stadt Bern, 1992, p.2) with the incorporation of all 'regionally relevant policy areas' (Gemeinderat der Stadt Bern, 1992, p.1). The main focus is on the establishment of two development areas on the periphery of the city centre the principal access to which will be by the extension of the existing rail network (Gemeinderat der Stadt Bern, 1992, p.15). This plan uses models from cantonal planning strategies and is based on the suburban rail concept among others. Future employment will only be created around this main rail link, and passenger vehicle traffic will be restricted. In criticizing existing transport plans from 1982 this urban development concept marks the end of a purely technical emissions-oriented environmental policy. Requirements contained in the Federal Air Pollution Decree and other clean-up measures described in the Federal Environmental Protection Act (*Umweltschutzgesetz*, USG) are integrated into plans known as 'accompanying environmental plans' which deal with all the sources of pollution that affect the city (air, green systems, global warming). With this new direction the city of Berne has entered what is known as 'ecological urban development policy' (Knoepfel, Imhof and Zimmermann, 1995, pp. 137–9).

Actor Profiles (Independent Variables)

The following section contains a description of the primary independent variables in the form of the most important actors and their institutional characteristics:

Municipal administrative actors

The transport inspectorate (*Verkehrsinspektorat*, VKI) is concerned with transport management. This includes all matters concerning the operation of transport infrastructure in the city, for example the control of traffic with the help of organizational and constructive measures, such as traffic lights or speed limits. The transport inspectorate has average personnel resources compared with other units of the local administration; its employees are mainly engineers. Due to the nature of its tasks it is generally highly interested in intervention in the field of transport; pressure from the public to legitimize its activities means, however, that this interest tends to be steered in varying directions. This body must deal with both requests to reduce the levels of traffic and requests for measures to improve the flow of traffic. Thus its interest in avoiding conflict must be as strong as its interest in the establishment of decisive transport policy.

The traffic division of the municipal police is responsible for the implementation of traffic regulations and checkpoints. It assumes part of the responsibility for the implementation of tasks on behalf of the transport inspectorate. However, it is not always possible to define its tasks so clearly. In reality tasks have tended to be divided between the two bodies on the basis of content, with the municipal police taking responsibility for stationary traffic and the transport inspectorate for mobile traffic. The institutional interest of the traffic police in interventions in the area of transport are as substantial as those of the transport inspectorate. The police must also pay attention to the capacity of measures to be implemented and controlled on a practical level, a factor which tends to undermine its interests in intervention. As the police are usually the last link in the implementation chain they see themselves as having to face all possible kinds of conflict with target groups – another factor which tends to dampen their interest in intervention and increases the pressure on them to legitimize their behaviour.

On the basis of the division of labour within the transport inspectorate, the Transport Planning Section of the Municipal Planning Office (*Verkehrsplanung im Stadtplanungsamt*, SPA) mainly deals with the development of conceptual and planning frameworks for measures to influence traffic in the city. In contrast with the system in other cities, the transport planning in this instance includes all forms of transport, for example public transport, bicycles and pedestrians as well as road traffic. This combined planning for all forms of transport gives rise to an institutional interest in interventions in the area of private passenger traffic. As the survey showed, this definition of the working area favoured a general approach to transport problems in the city of Berne and this in turn led to a predominantly 'ecological' stance on the part of the actors.

The remaining urban administrative actors (municipal works department, roads inspectorate) are mainly active in the areas of construction and technical implementation of measures. Their institutional interests mainly focus on the efficient implementation of measures. Due to the fact that they are not involved in the conception of measures, they are not under significant pressure to justify their activities. The involvement of the Municipal Transport Company (*Städtische Verkehrsbetriebe*, SVB) in the introduction of measures for the reduction of traffic levels is minimal. But the SVB, which has a large staff at its disposal, is without doubt the central actor in the measures to promote public transport. The main focus of interest in such measures lies in obtaining political guarantees for the subsidies and the optimum distribution of services.

The environmental activities of the Municipal Office for Environmental Protection and Inspection of Foodstuffs (*Städtisches Amt für Umweltschutz und Lebensmittelkontrolle*, AfUL) include the implementation of environmental impact assessments for construction projects, air-pollution control, noise-pollution control and the maintenance of water quality in areas surrounding industrial plants. The AfUL also acts as a general coordinating body for environmental issues and provides technical information on environmental issues. The AfUL's involvement in the projects studied mainly centred on the implementation of air-quality tests. Moreover, in conjunction with the environmental protection section of the Cantonal Office for Industry, Commerce and Labour (*Kantonales Amt für Industrie, Gewerbe und Arbeit*, KIGA), the AfUL also ensured the integration of measures to reduce traffic in the cantonal measurement plan as prescribed under the terms of the Federal Air Pollution Decree.

Regional and cantonal actors

Regional cooperation between the local authorities is institutionalized in the form of the Association for Cooperation in the Berne Region (*Verein für die Zusammenarbeit in der Region Bern*, VZRB). This association was hitherto mainly involved in planning issues and during the 1970s was responsible for various development plans. The VZRB is in a weak position as it does not have any absolute authority; moreover, its standing is disputed by some of the local authorities in the agglomerations. Of the four measures studied the VZRB was only involved in the collective fare agreement, for which it assumed the management of the project.

The main cantonal actor in the area of measures to reduce traffic levels is the transport technology section of the Road Transport and Shipping Office (*Strassenverkehrs- und Schiffahrtsamt*, SVSA). According to the cantonal Road Transport Act the state is seen as responsible for public road signs and markings; the local authorities only require cantonal authorization to impose major transport restrictions. As a technical authority the SVSA is highly specialized and is under minimal pressure from target groups and 'victims' to legitimize its activities Its main institutional interest lies in the technically correct and consistent implementation of road signs in all areas.

The Department for Construction, Transport and Energy (*Direktion für Bau, Verkehr und Energie*, BVE) and its subordinate Public Transport Office (*Amt für offentlichen Verkehr*, AOV) are the cantonal instances responsible for the tendering, planning and financing of public transport. These bodies both occupy a leading position in the administrative hierarchy. In contrast to the general department, the AÖV is a highly specialized body and owing to the political attention focused on public transport it is under significant pressure to justify its activities to the target groups of measures introduced. Its institutional interests span the areas of finance and transport policy. In addition to the cantonal actors the transport companies should also be mentioned here. Of these the Bern-Solothurn Regional Transport Company (*Regionalverkehr Bern-Solothurn*, RBS) and, possibly, the Swiss Federal Railways (*Schweizerische Bundesbahnen*, SBB) are the most important. The institutional characteristics of both these organizations are broadly equivalent to those of the municipal transport companies.

Social actors

As the social actors share a wide range of characteristics they will be dealt with in the following section as a group. Hierarchical positions and degree of specialization are irrelevant here. The scope for action available to these groups depends mainly on the social group they represent.[3] The cycling association 'IG Velo' and the Berne regional group of the Swiss Transport Club (*Verkehrsclub der Schweiz*, VCS), which is devoted to the promotion of environment-friendly transport policy, represent one end of the spectrum. In addition to representing the interests of the 'victims' of transport problems these organizations mainly represent general interests. Their organizational resources are extremely basic.

At the other end of the spectrum we have the Berne division of the Association of Industry and Commerce (*Handels- und Industrieverein*, HIV) and local divisions of the automobile associations – the Swiss Automobile Club (*Automobilclub der Schweiz*, ACS) and the Swiss Touring Club (*Touringclub der Schweiz*, TCS). The latter mainly represent the specific interests of their members and have significant organizational resources at their disposal in the form of professional secretariats. They also have a considerable mobilization potential and corresponding access to the political authorities. It is more difficult to make an unequivocal statement on their attitude to traffic reduction measures than would be assumed on the basis of their *raison d'être* as associations; these attitudes depend to a great extent on the expected effect of the proposed measures on the groups they represent. The neighbourhood associations take a more central position as they represent the interests of both local inhabitants and commercial concerns. In Swiss cities these neighbourhood associations play an important role in local politics. In general they have a long history and their representative power is quite important, even in local elections.

AN EXAMPLE OF THE EMPIRICAL FINDINGS: TEMPO 30 SPEED LIMIT ZONES

The following section presents the genesis of the 'Tempo 30' speed limit zones and the resulting empirical findings as an example. The results of the other three measures or objects are dealt with in summary in the conclusion which also contains a table with an overview of the most important features (see: Table 8.1).

Early history of the Tempo 30 zones (1989–94)

In terms of perimeter and scope of intervention the introduction of zones with a maximum speed limit of 30 km per hour (Tempo 30) represents the most comprehensive measure introduced to reduce traffic levels in the city of Berne as it was conceived as a general measure to be operated in all major residential areas. As with the other measures examined, the final introduction of Tempo 30 zones represented the culmination of a lengthy preliminary process which ended with the city council's decision to allocate loans for exploratory projects in all Tempo 30 zones. This decision, which dates back to 1992, marked the end of the first phase. This was followed by the actual realization of the measures in individual residential areas; the Länggass neighbourhood will be examined below as an example.

Unlike the measure to introduce resident parking permits, the introduction of Tempo 30 was accompanied by intensive political debates. There were a series of preliminary interventions at parliamentary level in the run-up to the implementation of this project. At the same time the different interest-group initiatives on transport policy increased the pressure on local authorities to resolve the traffic problems. Considerable pressure could be observed throughout the city in favour of the implementation of such measures in residential areas; this pressure was also heightened by the emerging air-pollution debate. The city authorities had been involved in the introduction of various measures to reduce traffic since the early 1980s. These were, however, purely experimental measures which were implemented in six selected residential areas (Planungs und Baudirektion der Stadt Bern, 1988, p. 9) and as a result of protests, the authorities subsequently abandoned these measures (Polizeidirektion der Stadt Bern, 1991, pp. 11).

As an alternative to localized construction measures for the reduction of traffic the idea of a general speed limit was then adopted from Germany. This new move necessitated the revision of Federal road transport legislation. When this had been accomplished in May 1989, Berne city council immediately allocated a loan for a pilot project involving the implementation of Tempo 30 in four residential areas. With the help of the pilot tests it was intended to define the additional construction measures required and establish whether Tempo 30 would achieve the desired effects. Following the basically successful completion of the pilot projects in the spring of 1991 the city council passed a general resolution on the introduction of Tempo 30 speed limit zones one year later. Only a few months after the passing of the general resolution the city government presented the council with a loan request for the implementation of a total of 44 potential Tempo 30 zones and an information concept. At a meeting on 27 August 1992 the measure was passed by a significant parliamentary majority with only minor alterations to the original plan.

The second phase in the realization of the Tempo 30 zones in the Länggass neighbourhood overlaps with the first phase of the general resolution. Following the general resolution the project was approved by the city council without further debate and work began in May 1993. The city government launched a simultaneous publicity campaign. The implementation of the measure was not, however, completed at this stage as the effects of the measures could only be estimated in the individual neighbourhoods and when the actual construction work began. Various details of the design of crossroads had to be adapted or altered.

Changes to the Output: First and Second Phases

Areas with a speed limit of 30 km/h are intended to provide greater safety and improved quality of life in residential areas. This measure is mainly aimed at non-local drivers who usually try to avoid traffic bottlenecks by travelling through residential areas. Combined with the construction measures involved this measure is significant in terms of the scope of intervention. It is above all the general scale of the implementation of Tempo 30 zones which made the city of Berne into a pioneer in this field. When examined in this light the problematic aspects of the measure are diminished. It combines safety, the reduction of noise pollution and, to an unquantifiable extent, a reduction in harmful emissions.

The changes made to the measure between the first and second phases (the passing of the general resolution and the implementation of the measure in the Länggass neighbourhood) were minimal. Besides the details concerning local measures on individual streets, which were corrected during construction as a result of the intervention of 'victims' (residents), the question of street layout and the extent of the construction measures were the main points of contention. The potential cost and the importance of avoiding lengthy objection procedures led to the decision to opt for temporary measures. These temporary measures had the additional advantage of flexibility which enabled their adaptation to specific needs without great difficulty. The city administration and the political authorities thus tried to influence the process through the concrete division of the output. A further difference consists in the 'general applicability' of the object: in contrast to individual measures, the sign-posting of the zones subjects entire neighbourhoods to a different traffic regime. This feature proved very helpful in dealing with the demands of the fragmented victims' groups, a fact which was acknowledged in the survey by one of the actors.

Empirical Findings of the Policy Networks

The network involved in the first phase of the realization of Tempo 30 was average in size in relation to the other examined measures. The central actors included the transport inspectorate and the municipal police department. The city government (*Gemeinderat*), which was involved with the political aspects of the measure, also played a decisive role. The social actors were involved at a significantly subordinate level and were excluded in part from the planning. This finding can also be explained by the political debates which took place on the question of the reduction of traffic. From this perspective the measure can primarily be seen as an answer provided by the city administration to the conflicts within the city council and among interest groups. In terms of centrality this network was also standard; the only significant gap in power existed between the social and administrative actors. The network was quite dense and the pattern of relationships did not necessarily follow the hierarchical model of implementation. The conflicts and coalitions were not present on an open level. The measure was guaranteed on an institutional level by the transport committee of which the directors of the relevant authorities are members. A 'traffic reduction working group', which had previously coordinated the implementation of construction projects for the reduction of traffic levels, already existed. This meant that it was possible to avoid the fragmentation of the network at the administrative level and reveals a certain break with the social actors.

Although the network withdrew to local level during the second phase, there was no significant change in its size. New actors at implementation level, for example the roads inspectorate and the municipal works department, replaced the actors previously involved in the planning of the measure. With regard to the social actors these groups were reduced to those actors involved in the Länggass neighbourhood. The degree of centrality increased irrespective of the circular arrangement. The technical organization of the project was mainly determined by the transport inspectorate and the traffic division of the municipal police. With the exception of the city government the actors involved in the implementation of the measure played a clearly subordinate role. An unacknowledged coalition between administrative actors had already been formed during the first phase; this coalition gained in strength during the realization phase. There was consensus at operative level within the city administration concerning the form to be taken by the measure, whereas the social actors protested on the streets during implementation. The degree of the fragmentation along the boundary between the administrative and social actors was more pronounced than in the first phase. The basis for the network remained virtually unchanged with the exception of the increased significance of the 'traffic reduction working group'.

The internal structure of the network reveals certain information about the interests of individual actors. In general the administrative actors were interested in implementing the measure without further delay. Another reason was the grouping together of the often diverse demands made by residents for the reduction of traffic in residential areas in the form of a single measure. The search for consensus in the contentious field of transport policy will have influenced not only the behaviour of the city government as a political authority but also that of the transport inspectorate and the police who were in direct contact with the 'victims'.

Peripheral Conditions and Programme

As with other measures the Tempo 30 zones were later regarded as measures for reducing air pollution, although traffic safety and noise were still the main issues during the implementation of the pilot projects. It was after this that air pollution was introduced into the political debate as a further justification for the introduction of the measure. The cantonal plan for measurement of air pollution refers in very general terms (under the heading of 'stemming the increase in traffic') to the adaptation of speed limits on cantonal and local-authority road networks. The main influence in this instance was the political pressure imposed both directly and through the city council and the transport initiatives. Thus, the activities of the 'victims' contributed towards the realization of the measure. The decision to opt for temporary construction measures and signposting of the zones shows that the implementation of this project was also strongly influenced by cost factors. This was due not only to the city's budget deficit but also to the general recession which was under way by this point.

CONCLUSION

Comparison of all Measures

A comparison of all four examined measures immediately reveals that both the parking stands for bicycles and the collective fare agreement can be identified as measures for the control of behaviour through stimulus (Table 8.1). In such cases the scope of intervention is generally minimal – a factor which also influences the evaluation of the output. Conversely the residents' parking permits and the Tempo 30 zones both involved the strict regulation of behaviour.

Secondly a latent reduction of the measures in the course of their realization tends to arise in terms of the dynamic features defined in the research concept, perimeter, scope of intervention and scope of the problem. Problems at 'operational level', in particular, led to a retrospective adjustment of the measures. The residents' parking permits scheme is exemplary in this respect as this project showed that it is impossible to provide a complete definition of the details on paper. The bicycle stand project also underwent latent reduction when it came to the question of the locations in which the measure was to be implemented.

General patterns can only be partially observed in the features of the networks (see Table 8.1). The size of the networks, the degree of centrality and the density showed strong variations. The dynamics of network development were equally unequivocal. A series of core actors, for example the partnership between the transport inspectorate and the traffic division of the municipal police, were observed as active during the entire phase of realization, whereas individual actors tended to come and go at the periphery. With the exception of the collective fare agreement, the networks became more dense in the course of the realization of the project at local level. The absence of regional actors, for example agglomeration authorities, is also conspicuous in all cases. As opposed to this the features of institutionalization, fragmentation and coalition formation are very much present. In all cases the networks have firm institutional foundations in internal administrative working groups or project organization.

Table 8.1 Overview of all measures examined in the city of Berne

Objects	Bicycle stands	Residents' parking permits	Tempo 30 zones	Collective fare agreement
Dependent variable: output*				
Output	average	extensive	extensive	average
Change	(reduced)	reduced	reduced	reduced
Intermediary variable: network characteristics				
Size	average	average/small	average	large
Change	(constant)	reduced	reduced	constant
Centrality	minimal	extensive/av.	average	min./average
Change	(constant)	reduced	constant	increased
Density	average	high/deep	average	high/average
Change	(constant)	reduced	increased	reduced
Fragmentation	minimal	minimal	minimal	average
Change	(constant)	reduced	increased	constant
Institutionalization	extensive	high	high	high/average
Change	(constant)	constant	constant	reduced
Coalitions	Against individual locations	Pro-coalition in the municipal administration	Municipal administration /residents	Consortium/ rejection by local authority
Explanatory variable(s): actor's interests (summarized) and programme				
Actors' interests	Successful implementation avoidance of conflict	Extensive concensus in administration	Avoided disruptions, successful implementation	Financing securing of positions
Influence of measurement plan and LRV	Political argument (weak)	Retrospective evaluation	Retrospective evaluation, subsidies	Political argument

* Evaluated on the basis of perimeter, problem scope, intervention

In the three transport objects the urban transport committee provided an important forum where the leading officials from the police, urban planning, and transport inspectorate as well as executive actors, such as the municipal works department or roads inspectorate were represented. In addition to this, administration-internal working groups also emerged to deal with specialized issues. Fragmentation was consequently minimal as the transport committee assumed an important coordinating function. In those instances where fragmentation occurred, it was limited to the boundary between the city administration and social actors or 'victims'. Similar results apply to the collective fare agreement with respect to the fragmentation between local authorities and the consortium. The coalitions in some instances occurred along the lines of fragmentation although, with the exception of the collective fare agreement, the coalitions were not very strong. In the case of Tempo 30 and the residents' parking permits the pro-coalition feeling throughout the urban administration was minimal. In the case of the bicycle stands a coalition centred around operative viewpoints formed in opposition to some of the individual locations. It was only in relation to the collective fare agreement that the coalition with the decisive participation of the canton played the role of a pro-coalition.

Two Hypotheses for Further Examination of Networks in Environmental Policy

The link between policy networks and the output produced needs more detailed examination. Negotiations are complex processes. There are only a few characteristics which are linked directly to the policy output or measures. One of these characteristics is the size of policy networks and the integration of actors belonging to different policy fields or state levels.

One old finding should now be corrected: the biggest is not necessarily the best. An overall integration of regional actors and different policy fields may lead to unproductive political struggles and a reduction of the scope measures. There is an optimum size of policy networks, which varies very probably with the nature of the measures.

The profiles of the actors involved shows the same pattern of interests as earlier studies do (Knoepfel, Imhof and Zimmermann, 1995, pp. 354–9; even Jänicke, 1978, pp. 24–5). The participation of interest groups can be helpful for the implementation process, if there are some strong pro-control interests. At the local level that could be for example, residents, organized in neighbourhood associations. It is also possible to group together different pro-control interests in order to build a pro-coalition. In the case of Berne the most successful way of justifying specific measures was by combining safety, noise reduction and air quality arguments. If we look at the interest groups on the different state levels (federal, cantonal and local), the most promising combination of interests for air control measures in the transport sector are at the local level.

NOTES

1. The full text of this study was published in German in the publication series of the Institut de Hautes Études en Administration Publique (IDHEAP); see Marek (1995).
2. Knoepfel, Imhof and Zimmermann (1995, pp. 29–36) defined the following categories: driving practices (speed limits, road design), transport organization (for example, management of parking

amenities, traffic management); improvement of the public transport system (infrastructure, fares, etc.); reduction of private transport (fiscal measures, promotion of car pools, restrictions on goods traffic, and so on); residual category (including reduction in the number of private car parks and support for pedestrian and bicycle traffic).

3. Jänicke (1978, p. 24) draws attention to the basic problem whereby commercial interests are frequently easier to mobilize than environmental interests which have a weak organizational and financial basis. For general information on this topic see Von Beyme, 1980, p. 85.

9. Conclusion: institution building for sustainable urban mobility policies

Peter Knoepfel, Wyn Grant and Anthony Perl

The European Union's COST Action 618 project sought to assess policy initiatives in institution building and information campaigns for urban air quality management. These efforts centred on limiting the growth of metropolitan traffic, which is the main source of urban atmospheric pollution in most major cities. While this volume focuses on the challenges and results of institution building for improved urban air quality, Action 618 has also produced a book that analyses information campaigns that have been conducted to combat air pollution[1].

As discussed in the introductory chapter, the researchers who examined institution building in this volume adopted a common research design focusing on the question: What factors enable policy networks to develop and implement feasible and effective control measures for urban air pollution? Rather than considering these policy networks only as ends in themselves, researchers also evaluated the networks' capacity to enhance urban air pollution control and abatement. In certain policy networks, investigators identified a set of core beliefs and common practices that produced efforts to enhance urban air quality. These initiatives succeeded in introducing improved air quality objectives into policy domains ranging from planning, environmental assessment, policing of parking and road use, traffic management and urban transportation infrastructure development.

In this concluding chapter, we interpret some of the findings presented in the seven preceding case studies of urban air quality initiatives. We assess the experience of four countries (Canada, France, Italy and Switzerland) in light of discussions that took place at a COST 618 workshop on institution building where these results were first presented, along with contributions from other countries (Hungary, Denmark, Great Britain, Spain and Greece). From these analyses, we now highlight the policy dynamics that have demonstrated effectiveness in facilitating urban air quality initiatives. Integrating air quality management and urban transportation objectives has been shown to occur following some combination of: innovation in policy problem definition, the renewal of policy communities, the building of new institutional frameworks, the introduction of new policy instruments, and the facilitation of learning processes. Our objective here is both to summarize some analytical insights regarding how air quality and transportation priorities were reconciled and to synthesize some suggestions for those seeking similar, or even more ambitious, results elsewhere. In summary, this conclusion seeks to fill the need of policy actors who will have to respond to the growing local (and global) damages caused by urban mobility.

DEFINING SUSTAINABLE URBAN MOBILITY

According to *Our Common Future*, the report of the World Commission on Environment and Development, sustainable development requires economic and social activity 'that meets the needs of the present without compromising the ability of future generations to meet their own needs' (World Commission on Environment and Development, 1987 p. 34). Such a vision of sustainable development aims to ensure that 'the exploitation of resources, the direction of investments, the orientation of technological development and institutional change, ... are all in harmony and enhance both current and future potential to meet human needs and aspirations' (ibid., p. 46). Following this formulation and the growing recognition of the transport sector's pivotal role in facilitating sustainable development, various efforts have been made to develop an understanding of sustainable transportation. Canada's Centre for Sustainable Transportation (1997, p. 2) offers a broad definition, which emphasizes meeting the basic access needs of individuals and societies in a safe, affordable, and efficient manner. Sustainable transportation would do this while limiting emissions and waste within the planet's ability to absorb them and minimizing the consumption of non-renewable resources and land, and the production of noise. The Swiss Agency for the Environment, Forests, and Landscape (1997) has put forward a more focused concept, which expresses mobility in terms of personal CO_2 budgets that need to be balanced in order to prevent climate change. Such an accounting system could be applied to other damaging effects of mobility.

In order to advance both urban public health and quality of life, sustainable mobility must recognize that the means of movement (vehicles, roads, parking space, public transport and pedestrian facilities) as well as clean air are finite and valuable resources. Realizing an urban region's economic and social potential requires managing both these resources effectively. This means distributing access and the use of transportation infrastructure in a way that balances the reasonable mobility of today's travellers against future mobility needs by internalizing the negative external effects of urban travel (for example, air pollution, noise, accidents). In economic terms, this appreciation of sustainable urban mobility calls for an optimal allocation of two scarce resources: space and clean air. Such allocation would be economically efficient if it created an equilibrium between the marginal cost of each urban trip and the price which each traveller (or shipper) were to pay.

But public policy must also address the equity issues that arise when managing essential resources with many characteristics of a public good (for example, open access, near-universal availability), typical of both urban mobility and clean air. Politically, the trade-offs between clean air and travel opportunities that are implied by sustainable urban mobility trigger a classical distributional conflict among present and future user groups. For a wide range of stakeholders, the significant changes in personal behaviour, corporate activity and the current value of urban assets (property) that must occur in the pursuit of sustainable urban mobility will only be accepted as legitimate when they are seen to be both efficient *and* fair. Some of the European and Canadian policy-makers who are profiled in this book have recognized that road traffic generates the bulk of urban air pollution and are leading the way toward managing air quality and mobility as interdependent resources. In other cases which we explore, the political legitimacy required to pursue truly sustainable urban mobility remains to be achieved.

Experience to date demonstrates that the equity issue underlying efforts to achieve sustainable urban mobility is crucial to gaining legitimacy. Each policy initiative that seeks to trade off current urban mobility patterns against enhanced air quality must demonstrate scrupulous fairness in respecting the rights of all parties, from frequent travellers to those who are particularly harmed by air pollution (such as the young, the elderly, and those who are ill). Institutions dealing with urban air pollution must be open and robust enough to control this high degree of distributional conflict while still guaranteeing positive and efficient allocation.

Integrating urban travel and air quality management policies is rarely, if ever, accepted as a zero-sum trade-off between drivers and citizens – if for no other reason than most citizens either travel by motor vehicle or aspire to do so. Citizens are wary of changing established urban travel behaviour in order to clear the air they breathe when the mobility alternatives appear unfamiliar and less attractive. Thus policy solutions must deliver a positive sum result of better urban mobility and cleaner air. This means transcending old political conflicts between automotive interests and mass transportation interests (many of which now reside in public bureaucracies). Non-motorized travel options including bicycle, pedestrian and telecommunication alternatives need to be brought into the sustainable urban mobility attainment scheme. Furthermore such efforts need to focus on a geographically complete zone of mobility, not just those trips which lie within urban political boundaries. This means dealing with local, metropolitan, regional and even national mobility patterns, as Lyon's predicament illustrates.

All contributions to this book underline the necessity of bridging what are today more or less isolated individual policies and their fragmented communities dealing with singular aspects of urban mobility systems such as road construction, traffic regulation, land use planning, or industrial and domestic environmental policies which address emissions that pollute the same airspace as automotive traffic. Each of these policies has a well defined mandate and a more or less exclusive policy community. But the current urban crises in public finances, growing congestion and increasing air pollution have opened a unique window of opportunity both for intensified cooperation across environmental and transport policymaking activities and for subsequent integration of a much broader urban mobility policy community including all those who have a stake in mobility and clear air. The Berne and Vancouver cases presented in this book, as well as the emergence of integrated urban traffic plans in France or Italy, show the way in which this development can evolve. We turn now to pinpointing the changes in problem definition, policy community composition, institutional frameworks and learning processes needed to create this new capacity for sustainable urban mobility.

A NEW PROBLEM DEFINITION

Going through the seven contributions to this book, it becomes evident that, once urban mobility reaches the point at which it becomes a cause of serious air pollution, a wider range of more or less visible social costs will have also arisen. As the Italian cases illustrate most clearly, some of this collateral damage may be perceived as more problematic, or more immediately troubling, than the generalized human health degradation, local and global environmental damages or decay of historical monuments, which arise from urban air pollution. Efforts are needed to fashion a problem definition of sustainable urban mobility that encompasses both the concentrated and imminent impacts of unsustainable travel patterns as well as the diffuse and cumulative effects. Such an amalgamation offers the best opportunity to mobilize support for what are inevitably contentious objectives and instruments. We offer seven components of this encompassing problem definition which can maximize the opportunity for policy innovation:

Congestion

Traffic congestion, a problem explicitly used by the Turin authorities as a more politically palatable proxy for managing air pollution, has been on both the Canadian and European policy agenda since the 1960s. For a long time, the congestion problem was addressed through supply side solutions of building more new roads, or self-contained public transport infrastructure such as underground railways. Such new infrastructure was supposed to reduce congestion either by increasing the physical capacity for traffic or shifting some part of urban travel from roads to subways. Newman and Kenworthy (1988) document how this approach was ultimately counterproductive because it neglected the effect of new infrastructure on decentralizing urban land use. More infrastructure stimulated longer trips because distant spaces became better linked together, yielding higher traffic volumes, along with greater energy use and air pollution. Lyon's postwar experience with supply-side infrastructure expansion clearly demonstrates the effects on sprawling land use, greater traffic volumes, renewed congestion, and increased air pollution.

By the 1990s, the effects of supply side congestion management policies had reached the financial and physical limits of many European cities. Space and money for the construction of new roads and new public transportation systems became increasingly scarce and urban authorities were compelled to consider new instruments that turned to the management of travel demand. Urban governments introduced various regulatory schemes, as well as more significant parking charges in urban areas as a way to limit the demand for road space. Marlot and Perl's chapter on Lyon shows that demand side management policies have even progressed to the point of experimenting with urban road tolls and consideration of more widespread road pricing or traceable use permits. In so far as demand management strategies reduce the absolute number of cars in use daily, they can also contribute to cutting down on air pollution (Perl and Han, 1996).

Air pollution

This 'core' problem was initially identified and championed from the political fringe. Across Europe, urban pollution has often been used as a political weapon of green parties and ecological movements. Along with noise pollution and road safety, the air pollution problem has become part of a relatively recent, but growing, critique of the automobile and its role in undermining the quality of urban life. This negative perspective on automobility usually spurs calls for restrictive command and control measures to restrict urban traffic – either geographically as in Turin's car free zones, or temporally as in the emergency limits placed on car use during episodes of peak air pollution found in France and Greece. Both the critique of the car and its accompanying restrictions have received little support from two important urban political constituencies. At one extreme, the urban working class still aspires to the automobile, and related material consumption which greens and ecologists denounce as unsustainable. At the other end of the spectrum, urban economic elites embrace car and truck transport as necessities for business. The most receptive social segment to this 'post-material' view of urban life has been the middle class, which is a declining segment of the population of many cities. When mobility and air quality management cross the urban boundary to include the outlying areas where many middle class voters have settled, air pollution is given a higher political priority based on quality of life concerns. For example, political campaigns targeting air pollution were found to have a significant impact in the outer belt of Swiss metropolitan areas (Knoepfel, Imhof and Zimmermann, 1995 pp. 145 and 365). But in many European cities where traditional political boundaries impeded the politics of air quality management, attention to air pollution often arose only once the need to implement European Union directives became pressing.

Safety

Efforts to enhance urban traffic safety, such as reducing speed limits in cities or traffic calming of roads in particular neighbourhoods, can complement sustainable urban mobility initiatives. The road safety issue has been a longstanding concern for both urban parents and senior citizens. As such, it offers another potential counterweight to the equity concerns that would arise from any efforts to limit or regulate urban mobility. When seen as an equitable sharing of risk, safety concerns also raise the opportunity to gain support for sustainable transportation policies from cyclists and pedestrians seeking a fair share of urban public road space.

Urban Fiscal Crisis

As shown above, supply side solutions to traffic congestion turned out to be very costly. When cities could afford massive infrastructure budgets, new highways and metros were seen as a mark of affluence. But as many well-to-do taxpayers have moved to outlying jurisdictions (often encouraged by the ease of travel that the new highways and subways created), municipal governments are being compelled to stop splurging on new infrastructure. Some jurisdictions have sought to raise revenue from infrastructure through road and bridge tolls and other user fees, while others have raised taxes. This revenue enhancement strategy offers limited room for fiscal manoeuvre owing to the public's low tolerance of new taxes. Taxpayers' revolts and the rise of various forms of protest parties demonstrate the limits on government in using pricing to support sustainable mobility. Turin's experience illustrates how politicians currently cope – by exempting specific population categories from the new tax obligations. New room for economic instruments might arise through the use of public private partnerships, where the tolls are collected by a private agency. Or governments may succeed in selling urban transportation pricing as a valuable new public service which enhances the quality of life by limiting congestion and pollution.

Discrimination

This issue is particularly delicate and usually implicit in mobility debates. When faced with the prospect of higher prices or restrictions on urban transport, some people will react by blaming 'foreigners' for the problem. To some, these outsiders are the commuters coming from low density areas, while others see recent immigrants (or visible minorities who are assumed to be recent immigrants) as causes of urban degradation. In either case, newcomers are seen as preventing 'legitimate' inhabitants from the full use of 'their' roadways, parking spaces, public places, or clean air. Rather than accepting lower standards in these areas, or changing behaviour, the temptation arises to try and exclude outsiders from urban roads and travel. The discrimination issue becomes even more contentious when drivers challenge other users of urban road space (pedestrians, cyclists, demonstrators, and so on) as being less legitimate; or vice versa.

The discrimination issue reveals different conceptions of property rights on specific roads and parking spaces. It pits those who are 'in' against those from outside and reinforces the typical fragmentation of municipal-based institutional arrangements in traffic, but also in fiscal, physical planning, and cultural policies within the metropolitan area. This dispute between insiders and outsiders takes on a critical dimension in the chapters on Turin and Lyon. When discriminatory policy solutions are adopted for achieving sustainable urban mobility, they create a risk of exacerbating social exclusion and widening the gulf between ghettos of the destitute and enclaves of affluence.

Equity

The equity issue is intimately linked with the discrimination issue. It usually arises in fragmented debates regarding the benefits of scarce urban mobility resources, such as parking spaces, road calming measures, permits for access to traffic restriction zones (Turin), new public transportation infrastructure, or the dedication of some existing road space to public transit through bus or tram lanes. The stakes consist of who will derive the most benefit from new or improved accessibility, whether by public transportation, enhanced parking, or traffic regulation. This issue becomes linked with air pollution when mobility management plans propose large scale realignments of accessibility, which can have serious negative impacts on the value of urban land.

Quality of Mobility

As Giuliani's chapter points out, the organization and delivery of urban mobility services can make a difference in the qualitative perceptions of travel alternatives. Public transport that offers seamless travel through, for example, integrated fares, coordinated timetables and accessible stations, can be an attractive alternative to the automobile. But costly, crowded, and limited transport facilities widen the quality gap between the car and its alternatives. Too often, quality measures and standards are disjointed, with one, usually high, set of expectations for automotive travel and another, usually low, set of expectations for all other transport options.

One can argue that, although each of the seven issues described above has its particular social or financial constituency, they belong together under a common framework of sustainable urban mobility. That is because the set of interests behind each interest are actually connected by their need to share a largely finite, and increasingly contested, urban transport infrastructure on the one hand, and their overall responsibility for the cumulative externalities of urban mobility on the other hand. The collective problem at the root of comprehensive urban mobility policies is how to organize urban mobility in a way that optimizes the internalization of social and environmental costs, and equitably distributes the burdens of achieving sustainability (including any remaining external costs) in a way that guarantees minimum mobility standards, minimizes damage (safety, clean air, noise, quality of life), and maximizes quality of life.

New Issues – New Policy Community

Public policy communities are nominally composed of state and societal organizations that share an interest in a common policy issue. Within these policy communities, actors will pursue varying dynamics of cooperation and conflict, to address the issues they share some stake in. Each organization brings its own intellectual and material resources to the community. In the case of sustainable urban mobility, these could range from the strongly held beliefs of radical ecologist groups to the analytical tools of scientists and social researchers to the money and employment opportunities controlled by private industry. One category of participants brings a unique resource to the policy community; state actors, whether bureaucrats or elected officials, have the legitimate use of authority at their disposal. This ability to compel certain actions, or to levy taxes or other mandatory fees, gives the public sector participants a decisive role in policy communities, which can take on quite different structures depending on how, if at all, this authority gets distributed. In most communities, one can identify subcommunities, where public and private organizations cluster together by function, by region, by ideology, or by other shared characteristics. These groupings are influenced by the levels of economic and political capacity among societal organizations in relation to their governmental counterparts. The common denominator that both connects these subcommunities, and orients their interaction within the larger community, is their organic connection to the definition of a public problem. Any definition of what is to be done in the name of the public interest, and with the attendant authority and resources of government behind it, will trigger the involvement of some organizations and not others. As the contributions to this book have demonstrated so well, what gets done about urban mobility and air quality depends on who participates in the public deliberations over objectives, instruments and institutional arrangements that fall under the definition of policy development. And that participation depends on how the problem is presented.

When urban mobility policy is approached from the understanding that traffic congestion is the key problem, we find little or no participation in the policy community by environmentalists, industries which generate emissions from stationary sources, community groups concerned with the safety of children or the elderly, public health agencies, and air pollution control agencies. Conversely, when road safety, urban revitalization, or the equitable access to public roads come to be understood as the key problems arising from urban mobility, then a very different set of actors will take part in the policy community. In this case, the elderly, homeowner associations, banks and financial agencies, even political parties representing particular segments of urban society, will participate in policy debate and development.

For example, the French urban mobility policy community shows a clear division into subgroups focusing on traffic engineering and air quality management. But in France, each reconsideration of the policy problem (say, from the lack of sufficient urban transport infrastructure to an inefficient use of existing facilities) brings new actors into the community. French experience shows that treating air pollution as a problem that is both equivalent to, and interdependent with, traffic congestion will inevitably enlarge the policy community membership with newly mobilized politicians, scientists, nongovernmental organizations, public transport agencies, and even private industry. Each of these actors is interested in having a say in the development of newly mandated urban mobility plans (plans de déplacements urbains; see Chapter 6, pp. 102). In general, linking all seven of the above mentioned problem definitions under the umbrella of a sustainable urban mobility policy will inevitably and considerably enlarge the policy community and render policy development more complex than it is today. Based on what our contributors have discovered in a wide range of urban mobility contexts, we can pinpoint the fulcrum on which a policy community capable of pursuing sustainable urban mobility could be balanced.

Respect for *property rights*, both the rights of private proprietors of the urban space which traffic passes through and the public trustees of commonly held resources such as clean air is the key concept that can engage the range of groups and organizations needed to achieve sustainability in concerted action.

Marlot and Perl's chapter echoes the finding of Ostrom et. al. (1994, p. 4) that achieving sustainable development demands a preoccupation with institutional economy in general, and property rights in particular. Ostrom's framework differentiates four categories of property rights ranging from the owner's freedom to unilaterally modify the resource in question to a circumscribed autonomy to use a resource in a particular manner, and up to a specific quantity based on terms set by some contract or regulatory code.

The legal definition of most transport infrastructure, particularly urban road infrastructure, assumes that this property is a public good whose use cannot be restricted. This public good assumption also implies that no one's use of urban road space will detract from any other use of that space. But this central premise upon which many urban transportation problems have been defined turns out to be defective. In actual fact, both traffic congestion and the exclusion of nonmotorized traffic, such as cyclists and pedestrians, demonstrate that road infrastructure is not truly a public good, but at best a mixed good, and in many cases even a private good being supplied at public expense. By accepting unrestricted 'first come, first served' automotive access for urban roads and enforcing the restriction, and even exclusion, of nonmotorized transport, the state implicitly advances the interests of certain drivers and landowners over other drivers and landowners, as well as cyclists and pedestrians. The same is true for clean air, even though the property right attributed to one user group (travellers using cars and shippers using trucks) to the detriment of others (inhabitants suffering from air pollution) is much less visible and more difficult to communicate. According to this very simplified formulation of institutional economy, sustainable mobility policies will have to define new and more realistic categories of rights for both clean air and road use. This redefinition will entail a major redistribution of these rights which is certain to be politically controversial.

If urban mobility was reconceived of as an amalgamation of more-or-less private property rights, with the attendant need to reconcile competing claims, rather than as a public good, then a new policy community composition and configuration would be inevitable. This community would be composed of both the *de facto* owners of property rights which currently govern urban road use and air pollution, as well as those citizens who are negatively affected by current rules of the road through air pollution impacts or restrictions on access to public roads. Among the policy actors who would face significantly changed roles and responsibilities in the new policy community, we find:

- *The state* itself through local, regional, or national governments functions as the primary owner of transportation infrastructure – roads, railways, parking spaces, transportation terminals, and so on. These functions are

normally managed by public road authorities, public works departments, or more or less independent road construction and operating agencies.
- *Private corporations* also supply urban transportation infrastructure through building and running facilities such as highways, tunnels, bridges, and even subways, light rail transit (LRT), and tramways (as in Hong Kong and Singapore). Here the full cost of infrastructure, including the cost of capital, is recouped from user fees and tolls.
- *Public transportation agencies* are responsible for operating urban transit systems. In the case of underground and tram systems, these agencies are the exclusive infrastructure operators, and in many cases public transportation agencies are also accorded a monopoly over common carrier passenger transport over city streets as well.
- *Private car owners* are the most common 'consumers' of urban road infrastructure and parking spaces. Car owners link political jurisdictions in the pursuit of urban mobility since the drivers who become part of urban traffic may come from the city, the metropolitan area, or even further away. These car owners are also the most numerous 'producers' of urban mobility externalities since the fuel and vehicle taxes they pay do not cover the full costs of noise, accidents, pollution, capital investment in road infrastructure, and the rest.
- *Public or private parking space owners* lease the right to store vehicles on prime urban land on various terms. The prices charged, and the distribution of parking supply, can have a decisive impact on the efficiency of urban public transit.
- *Urban landowners* more generally have a major stake in the terms of urban mobility. Easily accessible urban subway commands an economic premium as the development around urban metro stations or suburban highway entrances and exits demonstrates. At the same time, transportation infrastructure that passes through, or near to, property without improving access can reduce its value by bringing externalities (such as, noise, fumes, and physical isolation from other property). From homeowners to big corporations and real estate holding companies, property owners pay close attention to transportation planning because of its influence on their assets;
- Local or regional *air quality management agencies* exercise public powers to limit air pollution through regulation, inspection and reporting on the state of the environment. For the most part, these agencies have been called upon to take stewardship of degraded air resources long after the public good concept had been entrenched as a cornerstone of urban mobility. Once the right to pollute excessively had been handed out to urban drivers, these agencies were given the unenviable task of reasserting public authority to limit the unequal distribution of damages (for example, protecting the

young and elderly who are less mobile and more harmed by air pollution). Faced with entrenched opposition from drivers who resist giving up their free ride, these agencies would be tempted to compromise by focusing their regulation on less formidable political adversaries including 'foreigners', newcomers, future generations and industry.

The transformation of today's disparate policy community environment, where air pollution and urban mobility problems are often addressed by disjointed policy initiatives, appears to depend upon reconsidering the first principles of economic rights and political responsibilities. Such change is unlikely to occur spontaneously, at least not in the first instance. Instead, the problem redefinition required to launch this transformative policy process will need to be championed by a vanguard of policy innovators. Below, we speculate on the groups and organizations with the most reason to advocate an encompassing policy community. These concentrated stakes in sustainable urban mobility may be enough to mobilize their leadership in such a vanguard.

The gains most likely to motivate a vanguard of advocates for sustainable urban mobility rights will arise when the redefinition of today's *de facto* and *de jure* user rights gets translated into a restructured relationship between urban mobility and pollution. This requires either withdrawing some rights from today's travellers or even destroying some portion of right to automobility as Marlot and Perl explore in Chapter 7. They introduce the standard of an ecological carrying capacity for a region's airshed as a principle for reconciling mobility with pollution. Working from this carrying capacity, one can translate a fixed quantity of motorized mobility, which can then be bid upon or rationed. Whichever strategy gets pursued, compensation payments can become a key means to achieving such wholesale change (Small, 1992). Until such redistribution occurs, there will be increasingly concentrated losses which can motivate certain groups and organizations to activism. Amongst these *affected* groups we would highlight:

- *Inner city pedestrians, cyclists, and drivers* are each in their own way excluded from mobility options as a result of the current treatment of urban roads as a public good. For each of these travellers, the influx of vehicles from surrounding areas leaves little opportunity to use adjacent road infrastructure. They seek new access to the public domain of urban mobility which could be achieved through either a new institutional economy of allocating mobility rights, or a simple restriction on the use of certain public space – pedestrian zones, protected pedestrian crossings, bicycle lanes, closing roads to through traffic, and so on.
- *Public transit operators* suffer badly from having their buses and trams caught in today's traffic congestion, and will suffer even more from the

decentralization and dispersion of population and activity that accompanies growing road infrastructure. Their best chance of offering improved quality to existing passengers and attracting new users comes from ending the 'free ride' policy for urban drivers;

- *'Auto-dependent' urban residents* have little choice but to drive or face sharply restricted mobility because they live in areas that are poorly served by public transport (Newman and Kenworthy, 1989). Their best hope for a fair share of access to the metropolitan region comes from making public transport a more effective alternative to the car, which depends on realigning mobility priorities;
- *'Excluded' drivers* coming from either less central urban neighbourhoods or the surrounding low density suburbs are often penalized by traffic restriction schemes which exempt inner city residents. The resulting hierarchy of access can leave residents from outside the city centre doubly excluded in that they are poorly served by public transport, yet blocked from using their cars to access cultural and commercial amenities as well as job opportunities. A more equitable distribution of urban mobility would have strong appeal to individuals who are caught between such restrictions;
- *Inner city residents* face elevated levels of air and noise pollution, which pose their most serious public health threats at high levels of concentration. As a result, they are excluded from enjoying a minimum standard of clean air and tranquillity because these resources are consumed by drivers, and other urban polluters (such as industry). Often there is a correlation between the degree of negative environmental impact and socio-economic status. Inner city dwellers thus have every incentive to bring the social component of sustainability into a redefinition of urban mobility problems;
- *Industry and other fixed source polluters* have the legal right to generate emissions into an urban airshed, within regulated levels. As the experience in Lyon and Vancouver demonstrates, these fixed source polluters identify an economic advantage in shifting the burden of emission reductions on to mobile sources, such as cars and trucks. This motivation is strongest in new industries which must obtain permits for additional emissions, for established polluters who are (often wrongly) accused of 'causing' regional air quality problems, and for sources (such as paper mills, cement plants and thermal power generating stations) which seek to expand output to meet growing demand for their outputs. In each case, reducing mobile source emissions could yield concentrated benefits, such as new jobs and higher profits for urban industry.

Having mentioned the groups and organizations that comprise potential vanguard for sustainable urban mobility, we must recognize that there are also organizations which would mobilize to oppose such an expansion of the problem definition and policy community. Among the groups that have traditionally resisted any change to the public good definition of urban road use and its consequences are small and moderately sized urban businesses (often represented by Chambers of Commerce or Boards of Trade). These retail merchants and service providers, who are often quite powerful in local politics, as seen in Lyon, fear a loss of business if the terms of mobility are revisited and current customers are affected. At the other extreme, multinational energy producers and suppliers tend to oppose efforts to reduce transport sector externalities, given its downstream impact on their markets. In between, there will be many other groups that see a broad-based urban mobility policy community as a threat, which is why the vanguard we have identified will have to pursue a strategy that can compensate some of the losses in any redistribution of urban mobility opportunities.

BUILDING NEW INSTITUTIONAL FRAMEWORKS ...

What institutional framework is capable of adequately governing such a sustainable urban mobility policy? When the experience of preceding chapters is factored into the organizational principles and dynamics for achieving a considerable redistribution and realignment of urban mobility outlined above, we see the need for an institutional framework of rules and norms that would fulfil a minimum of four conditions.

First, there must be a *multi-stakeholder forum* composed of pre-existing or new organizations that take responsibility for five key policy components of sustainable urban mobility. These include: public roads (ownership, construction, maintenance and management of transport infrastructure); road police (regulation of access and rules of the road); land use planning (integration of transport infrastructure into the overall scheme of urban development); public transportation planning (development and management of public transportation infrastructure and operation); and, environmental agencies (management of natural resources including clean air, water and soil.)

These five agencies would require an appropriate incentive structure to work cooperatively in implementing a political commitment to achieving sustainable mobility. They must contribute their unique administrative and analytical capacities to this common goal. Cooperation should not be confused with organizational amalgamation into one huge mobility management agency. Such an agency would not only be unwieldy, but also threatening to the existing bureaucracy charged with managing various pieces of the urban mobility puzzle. Instead, the existing competency and experience of today's *de facto* urban mobility managers are valuable administrative assets that would be further enhanced through joint action, rather than bureaucratic rivalry. The ability to integrate sustainable urban mobility objectives into related, but otherwise autonomous, bureaucracies responsible for urban development (land use agency), natural resource management (water management influenced by major transport infrastructure, green spaces and so on) or public works agencies (public buildings) appears far more likely through joint action than attempts at formal amalgamation.

A second organizing principle for institutional arrangements ought to be their *trans-jurisdictional structure*. Our contributors have shown that urban mobility management cannot be limited to the administrative jurisdiction of the central city, but must be extended to the whole metropolitan area. The experiences of both Lyon (Communauté urbaine Lyonnaise) and Vancouver illustrate how the political debates triggered by congestion and pollution impacts, the equitable distribution of mobility, and the discrimination issue can be addressed most effectively within a jurisdictional framework which is enlarged to include the suburban municipalities. This transjurisdictional structure must bridge not only governments, but also private organizations and societal groups throughout the metropolitan area. As with the coordination of multiple stakeholders discussed above, crossing jurisdicational boundaries should not imply amalgamation or integration into regional bodies. The cumulative policy capacity of organized actors throughout a region will be greater if existing organizations work together, rather than be reformulated into a new structure. The Italian and French cases show that the range of jurisdictions can stretch to include regional government (Rhône-Alpes), national government (France and Italy), and even supranational government (European Union).

A sustainable urban mobility initiative will need to adopt our third organizing principle of *multipurpose political-administrative arrangements*. Although policy development should pursue the single and overarching objective of sustainable urban mobility, this goal covers more than one traditional policy issue. As we pointed out above, at least seven controversial issues fall under this objective; each of these is associated with a specialized knowledge base, which can create a varying number and configuration of policy subcommunities. These issues might thus be considered as a sub-goal of the overall objective and then be assigned to a specific policy actor within the coalition or partnership. In general, sustainable urban mobility initiatives will have to develop the level of trust and recognition necessary for such a division of labour and specialization of tasks.

Finally, the great diversity of policy participants, their range across multiple jurisdictions, and their reliance on specialized functions will all require a high degree of *horizontal and vertical integration* complemented by necessary vertical centralization. The utility of highly integrated political and administrative arrangements is demonstrated in most of the contributions to this book. For example, Giuliani finds that frequent intense contacts between institutional actors seems to be a precondition for coherent policy outputs. We would go even further and suggest that these contacts must not only be vertical, but also horizontal, linking organizations of the same juridical level from public agencies to industry councils in pursuit of the common objective. Whether these contacts are more formal or informal will matter less than their frequency and substance. We agree that frequent intensive contacts will make the arrangement more complex than arrangements of less comprehensive policy domains, such as previously segmented environmental and transportation policies.

Probably the most controversial characteristic that we are proposing is a centralization of the necessary authority needed to empower the many actors and organizations identified above to pursue sustainable urban mobility. We are convinced that even when the most effective and sophisticated policy instruments are used, as discussed below, the redistribution of user rights to clean air and mobility will be highly controversial and politically contentious. In such circumstances, innovation will only succeed if policy actors have the institutionalized authority to discriminate between competing private interests in pursuit of the public interest. An example of such a decision would be compelling a highly mobilized and politically powerful group or organization to give up some share of current urban mobility and/or pay for the damages created by that mobility. This task will require some politically powerful central actors in both metropolitan and urban governments. These actors should exercise their authority through the legal power to tax transportation activities, to regulate land use, and to build (and in some cases dismantle) transport infrastructure. Their powers must be legitimized through democratic elections, both of individuals to the appropriate public office and by the passing of referendums on the policy framework required to implement sustainable urban mobility (for example, full cost pricing of externalities, infrastructure redevelopment programme, and so on). The Italian and Swiss cases demonstrate the power of direct democracy to break down entrenched obstacles to redistributive reform, although Grant's (1996) review of how direct democracy is practiced in California sounds a cautionary note. In the final analysis, we believe that hard policies do require hard institutional arrangements to translate the ideas and options generated by the soft institutions that Giuliani presents into concrete actions.

The sustainable urban mobility policies we are considering do not yet exist. But the contributions to this book reveal some important insights regarding the preconditions of such policies. Breaking through institutionalized barriers to sustainability, which is what the cases profiled here have attempted, is a necessary start to the transition. We would expect that each effort to refashion the disjointed, and often adversarial, policy efforts in air quality management and urban transportation will have to pass through a phase of experimentation akin to those described in preceding chapters. But these initiatives represent only the first stage of a more profound transition to sustainability.

While new ideas can occur spontaneously in the interactions of an encompassing policy community, and while these new ideas can reshape organized interests, there remains a need to institutionalize new patterns of authority to achieve progress toward sustainability. Thus, after problem definitions broaden to correspond more closely to sustainable development, and policy networks are sufficiently in flux to accommodate new patterns of group interaction, there remains a crucial need for institutionalizing the emerging understanding of what needs to be done into a set of organized obligations. By this, we mean conferring the legal and political powers to pursue policy initiatives on one or more organizations, which then gain autonomy within the policy network. We are convinced that these key institutional actors will have to play a dominant role in delivering policy outputs, both through their own initiative and through structuring the activity of other groups in the policy network. Vancouver's experience comes the closest in this book to demonstrating a critical mass of institutionalized authority in action, in the form of the GVRD. But achieving the type of redistribution of opportunity for sustainable urban mobility will require even greater levels of institutionalized authority, which we see as arising through a more widespread and visible democratic legitimation of future lead organizations.

The policy community literature is not clear about how and when organizations become leaders in implementing new paradigms like sustainable urban mobility. Some authors (Lembruch and Schmitter, 1982) portray cooperative dynamics, such as corporatism, as naturally emerging when policy communities are given the political space to fill the organizational space between hierarchies and markets.[1] Indeed, the intellectual antecedents of policy community analysis arose from the study of corporatism as a mode of policy development (March and Olsen, 1989). In our view, the initially loose, voluntary and informal associational links that arise in an encompassing policy community are necessary, but not sufficient, to achieve sustainable urban mobility. To realize their potential, these new relationships will have to become institutionalized, legitimated, and autonomous so that key actors can take decisive action without the necessary approval of all policy community members. In this sense, the suggestion that soft institutions can truly resolve hard problems is chimerical because the deliberations and negotiations that are implied in the concept actually become counterproductive if left to continue indefinitely. In such cases, certain groups inevitably come to oppose initiatives that were developed collectively, and work to resist or undermine subsequent implementation efforts. In sum, the soft institutions needed to clear obstacles toward sustainable urban mobility are a necessary step on the path of policy development, but they cannot achieve the end in themselves.

... CAPABLE OF USING NEW SETS OF POLICY INSTRUMENTS

It is important to understand that redistributional policy objectives like sustainable urban mobility cannot be achieved solely by the use of economic instruments. As important as tools like road pricing, energy taxation, and parking charges turn out to be in sending clear signals regarding the cost of urban driving, they must be complemented by other instruments that influence the social and the physical dimensions of urban life. These complementary policy tools include traditional command and control instruments such as land use planning (zoning) or prohibitions of car traffic in specific areas (for example, the restricted traffic zones of Turin) that can be used in more or less draconian redistribution of urban space from drivers to other users. New infrastructure development and finance is another longstanding instrument of creating new transport capacity, which can be targeted to maximizing sustainable mobility. Finally, persuasion and public communication can make inroads into seemingly habitual travel behaviour. The challenge is to discover a mix of all these instruments that optimizes efficiency, effectiveness, and equity.

Both Bobbio and Zeppetella (Chapter 5) and Desideri and Lewanski (Chapter 4) are probably right when they postulate that regulatory instruments will encounter a much broader and more militant opposition than either infrastructure expansion or economic instruments such as parking charges. This is because restrictive regulations have a more obvious and far-reaching redistributional effect to mobilize potential losers. Economic instruments which enable citizens to buy temporally or geographically specific user rights for public roads or parking space will be more acceptable than their complete reattribution to another category of the urban population (for example, pedestrian zones). But pricing roads and parking at levels that keep emissions within the carrying capacity of a region's airshed might be charge enough to cause secondary redistributional effects in terms of social discrimination. Prohibitive mobility pricing could thus lead to adversary mobilization just as easily as draconian regulation. Studies from Switzerland demonstrate that even the soft instrument of clean air campaigns which raise the issue of redistributing mobility can trigger political opposition to this part of their message (Zimmermann et al., 1997, p. Xll).

One way to assess, and predict, the varying degrees of political mobilization that could accompany different types of policy instruments is to estimate the degrees of freedom and constraints that they would create. For example, the attribution of user rights through regulatory policies that create 'car free zones' yields a high level of freedom for new recipients (bicyclists and pedestrians), but this corresponds to a zero-sum imposition of constraints on drivers.

Regulatory instruments thus appear extremely discriminatory in that they yield a high degree of constraints on the one hand and great liberties on the other one. In the historic centre of Turin, traffic restriction zones guaranteed both pedestrians and those drivers with special permits exclusive user rights to the detriment of drivers coming from outside.

Regulation's concentration of liberating and constraining impacts, which has often been seen as necessary to create minimum health and safety standards for vulnerable groups, is partially avoidable through the use of economic instruments. Because tolls, traceable permits and other pricing measures only become exclusionary when the target groups are not capable of paying, they offer the opportunity to change mobility patterns, and reduce those with significant negative externalities, while preserving individuals' freedom of choice. Even when pricing schemes restrict total mobility – during rush hours, or air pollution episodes, or around the clock for a given space, for example, they do so without specifying exactly who will be excluded. This makes them more publicly attractive than regulatory approaches.

Economic instruments thus belong in the forefront of sustainable urban mobility, although their pricing parameters might be calibrated based on the same threshold that would inform more coercive regulations, such as the carrying capacity of a regional airshed. As long as air pollution remained within acceptable levels, then more coercive instruments like regulation could be left out of implementation efforts. But during smog episodes or similar environmental emergencies, economic instruments would need to be supplemented by regulatory controls. A permissive complement to pricing instruments would be construction of new infrastructure and redevelopment of existing facilities. Such measures bring popular additions to urban mobility capacity, which if properly planned would yield positive sum gains. These improvements could range from low-cost enhancements to existing public transit facilities (coordinated fare and timetable systems) to incremental infrastructure expansion (new interconnections between existing transport lines or park-and-ride facilities for drivers) to new tram and subway lines. The cost of such efforts could be met by a part of the revenue raised from pricing instruments. Persuasive instruments such as advocacy campaigns for green transportation and information programmes about urban travel alternatives can also be considered permissive, and achieve greatest effectiveness when they target specific user groups for travel behaviour modification. Funding such campaigns from pricing instrument revenues would be unlikely to generate much controversy.

Politicians often believe that the cost of policy implementation will vary inversely with the degree of authority found in the policy instruments. That is, less authoritative policy instruments like persuasion and pricing will cost more to administer than restrictive regulations. The Italian contributions to this book demonstrate that this is probably a mistaken point of view. Regulation cannot be counted upon to 'speak for itself' and the cost of policing restrictions on road or parking access can only be financed in part by raising fees.[2] But collecting user payments also has a price, particularly when new technology and personnel will be needed to deploy sophisticated marginal cost pricing systems like peak hour road or parking fees. These costs do not disappear if fees are collected by the private sector, since public revenue would be forgone while supervisory and planning activities continue. Turin's experience suggests that private partnership in enforcement or pricing can reduce the legitimacy of such measures. The Swiss case also shows that campaigning for sustainable mobility has both a political and an economic price (Zimmerman, et. al., 1997, pp. 67 and 129). For all these reasons, we conclude that a cost-benefit analysis of the four instrument categories would reveal a far narrower spread in instrument costs than people expect.

Our case studies also reveal that legal constraints on local and metropolitan governments will be a key variable in instrument choice. These constraints normally are found in national or even in EC regulations which local governments cannot ignore without being sued by affected interest groups. The discretion available in crafting local urban mobility policies is currently limited by national road codes stipulating general rules on speed and access for highways, regional roads, and sometimes even local roads. Other traffic and parking restrictions often conflict with higher authority. National and subnational governments are even more jealous of the fiscal prerogatives associated with mobility. National regulations on car taxation, car inspections, and technical requirements for cars can each inhibit local initiatives. The same is true for local governments' autonomy with regard to different forms of intergovernmental cooperation, especially attempts to cross jurisdictional boundaries across a region or metropolitan area. Experience demonstrates that the further local policies move away from a classical regulatory approach towards economic instruments, the more they encounter provincial, regional or national governments' opposition to 'discriminating' in the treatment of all citizens in the jurisdiction in question or, in the case of the European Union, of attempting to restrict people's freedom of mobility. As a result, the field of action is usually wider for local governments using persuasive instruments like campaigning or traditional regulatory instruments than it is for the economic instruments which could make such an important contribution to sustainable urban mobility.

Finally, the policy instruments vary also with regard to their possible adaptation to specific urban districts or target groups. Car drivers belong to the famous group of 'inaccessible target groups' (Bressers and Ligteringen, 1997). For political, social, and economic reasons targeting must be considered as a highly difficult task in this field. The Italian cases demonstrate how both regulatory and economic instruments had to be retargeted after their introduction by means of special permits or exemptions. Studies from Great Britain and Switzerland demonstrate that targeting can make a major contribution to the success of air pollution campaigns (Zimmermann, et al., 1997; Schroll et. al., 1998). The highest need for a very sophisticated targeting strategy exists in the field of organizational and infrastructural development instruments. Here targeting is always associated with more or less participatory procedures opening up decision-making to the concerned local populations (without, however, giving up completely the autonomy of public authorities for their own political choices). For several historical reasons, targeting will be a less complex task in the field of traditional road policy measures such as the creation of pedestrian districts, which are shaped by well established administrative routines and procedures. On the other hand, targeting will be extremely difficult in the application of economic instruments because of their intimate ties with social policy and equity. Political compromises become inevitable in this very nasty business of potentially highly discriminatory decision-making.

The institutional framework of the proposed sustainable urban mobility policies must be simultaneously flexible enough and rigid enough to implement the whole range of the possible instruments simultaneously. The targeting of instruments needed to optimize sustainable urban mobility will require political backing that is strong and uncompromising, but also flexible enough to obtain the minimum necessary acceptance of policy outputs by target and affected groups. Negotiating procedures which facilitate problem-solving will be needed to ward off strategic bargaining (Scharpf, 1993). The institutional framework must contain sufficient financial, personal, and legal resources to implement more or less costly instruments. Furthermore, the need for discriminatory decisions against strong target groups must guarantee that the framework is capable of mobilizing the support of sufficient affected groups by means of additional political legitimacy through direct and indirect democratic procedures (direct election of metropolitan area authorities, referendums). And last, but not least, the metropolitan area policy framework must mobilize support from national government and international bodies in order to efficiently use all potential instruments and not be hampered by regional or central government constraints.

... AND TO FACILITATE POLICY LEARNING

Given the intellectual orientation of network, sustainability and public management literature, it is not surprising that all contributors bring up the issue of learning in their conclusions. Learning is coupled with 'soft instruments' in various ways, but three examples will illustrate the range of possibilities identified by our authors. For Desideri and Lewanski, learning becomes a policy output that arises when policy communities implement persuasive policies and information campaigns. For Giuliani, the interorganizational learning that occurs in encompassing policy communities which are horizontally coordinated by voluntary and spontaneous decision-making creates an important new policy input. And Bobbio and Zeppetella depict the interaction within policy communities as an ideal place for groups and organizations to discover and refine their economic self-interest in relation to other policy actors.

In each of these perspectives, the authors express their conviction that ideas and knowledge matter for the promotion of policies.[3] While this emphasis might seem to be at odds with our proposition (made earlier in this chapter) that sustainable development policy requires a strong dose of centralized political authority, the two insights are actually quite compatible. While the inevitably redistributive character of sustainability initiatives will require an institutionalized autonomy for implementation, this power needs to rest on a solid and legitimate democratic foundation. Obtaining that legitimacy calls for making sustainable urban mobility an overt political issue, both through referendums on key principles of sustainable mobility (for example, user pay schemes such as marginal cost mobility pricing) and in campaigns for elected representatives who will carry through such programmes. The primary legitimacy thus given to sustainable urban mobility will require all three of the learning modes outlined by our authors. These lessons will help carry the redistributional struggle through the rough and tumble of local politics.

Without active learning among both policy proponents and the urban electorate, we doubt that even a majority of urban voters could impose truly sustainable policy options through episodic democratic 'victories' at election time. If the relatively affluent and empowered segment of the population who perceive strong benefits and limited alternatives to currently unsustainable urban transportation arrangements votes overwhelmingly in favour of change, strong minority target groups would subsequently reject this democratic redistribution of mobility options as an infringement on their rights. Court challenges, and widespread noncompliance could be expected under such circumstances. Instead, the Italian referendum experience demonstrates that political campaigns must contain an effective element of targeted persuasion at key constituencies that will be impacted by the results.

Changes in beliefs and changes in behaviour are each correlated with learning, but the causal process is hardly straightforward. Empirically, the collective cognitive processes leading to new knowledge, new values, new problem definitions and, finally, to a new 'worldview' remain virtually unidentifiable. Scholars agree that social learning is a collective process embodied in complex interactions between all kinds of institutional and societal actors consciously exchanging resources in the interest of solving of a common problem. Such exchanges are not solely intentional, such as in learning from mistakes or even proceeding by 'trial and error'. Literature suggests a distinction between learning forms can help explain the range of different results. Educational efforts can range from mandatory education efforts (such as military, corporate or government training programmes) through instrumental or conditional learning patterns (for example, 'need to know' initiatives that enable specific activities) to more comprehensive efforts (such as systematic investigation and analysis) (Kissling-Näf, and Knoepfel,1998). When new polices get implemented, each mode of learning can help explain and legitimize change. Most of these variants can be found in our case studies.

Mandatory education is typical of policy communities that are hierarchically organized, such as when the Italian courts decreed that certain transportation options must be pursued, or when officials in the Rhône-Alpes region of France approved a new express highway. Here, local policy communities are simply taught how to apply, or cease applying, a particular policy instrument without understanding the reasoning behind its use. Instrumental learning is typical of air quality management policy-making. Through both regulatory and economic incentive approaches, polluters learn to change their behaviour as, and only when, they encounter a specific restriction or charge. Turin's development of traffic restrictions follows this instrumental approach; the city's authorities adopted a trial and error mode in adapting their traffic restrictions in response to negative side effects and unintended consequences. Eventually they 'learned' to use an incentive system of parking charges to achieve the desired objectives, including air quality improvement. Learning by imitation is also common in the more technical side of air quality management, as local authorities survey the world and try to copy the best available solution, which they identify with success in other cities.

Policy studies can contribute to the understanding of social learning by identifying specific behavioural changes which result from cognitive changes among key policy actors. Changes can be attributed to shifts in the belief structure through the perception, and adoption of new values. But changes can also arise from shifts that have a more instrumental logic, such as the choice among different policy instruments or short-term policy goals (Kissling-Näf and Knoepfel, 1998; Sabatier and Jenkins-Smith, 1993; pp. 13–39). Policy studies can connect both changing core beliefs and shifting tactics to variation in the make-up of policy communities and/or political administrative arrangements. Such participatory and structural changes are one indicator of ongoing learning processes. Another such indicator of social learning processes is a considerable change in policy outputs, especially when it yields more effective policy outcomes.

All contributions to this book dealt with ongoing or imminent changes within policy networks arguing that the process of interconnecting previously insulated networks in urban traffic and air pollution management can be considered to result from learning processes among policy actors. Some contributions also claim to identify changes in the commonly shared values of the actors moving from a supply side towards a demand side approach to management. The evidence presented from Berne and the Greater Vancouver Regional District reveals that several outputs of urban traffic policy became more environmentally sound because of changes in the policy community composition and in the actors' values.

Though still a fuzzy concept at this point, organizational and interorganizational learning will play an important role in sustainable urban mobility policies, which intentionally aim at generating conscious cognitive and behavioural changes in various target groups. Such learning will have to generate commonly shared frames of reference regarding urban transport and the environment, which will differ significantly from today's urban mobility paradigm. In this transition, people want legitimately to know where urban travel policy is heading before they commit to such change. We probably underestimate this need for security in redistributional processes as radical as the ones proposed by sustainable urban mobility policies. As a car driver, you only agree to learn about giving up your more or less vital mobility rights if others equally affected do the same, and if you know what you will get in exchange. It is therefore absolutely necessary that the new policy comprehensively defines the different categories of those user rights that both target and affected groups will be able to enjoy tomorrow as well as the concrete functioning of the future system. This is the most important precondition for initiating learning processes and moving policy communities towards sustainable metropolitan urban mobility policies. Despite the difficulty of this educational challenge, effective social learning will yield greater safety, more equality, and less physical and psychological struggle than the present daily civil war on our public roads and spaces.

NOTES

1. The fuzzy definitions of 'networks' both as prescriptive and descriptive analytical dimension makes the use of the term often very unprecise. It is evident that the network concept is a useful tool for describing interactive systems of actors independently of whether these actors are public institutional or societal ones. In this sense, the described policy community will clearly form a policy network composed of social and institutional actors. The above used term of political administrative arrangement refers to that part of the networks which of composed of institutional actors. The network metaphors then can be used for the description of both interaction modes within this arrangement and between actors of the arrangement and its societal partner actors on the level of social actors. Used in this purely descriptive way, the network notion is an interesting analytical tool. But this use prohibits specific qualifications as defining characteristics of networks in opposition to more hierarchical or more market-oriented coordination systems such as 'absence of a center', or 'horizontal instead of vertical coordination', for example.
2. The city of Rotterdam recently demonstrated that good planning of control activities may use considerable public income and that the number of fees depends largely on the number of policemen. The more policemen a city engages, the more it can increase incomes from fees.
3. For further elaboration on the cognitive or constructive approach advanced in this book, see the growing literature on policy learning such as Sabatier and Jenkins-Smith (1993) Kissling-Näf and Knoepfel (1998), Knoepfel et al. (1997) and Nullmeier (1993).

Bibliography

Allegrini, L. (1994), 'Degrado Aria, una Prioriti', *Il Sole 24 Ore*, February 28.
Anderson, William P. and Clarence Woudsma (1996), *Urban Transportation. Energy, and Air Pollution: A Comparison of Policy in the United States and Canada*, Hamilton: McMaster Institute for Energy Studies.
ARA Consulting Group, Inc (1994), *Clean Air Benefits and Costs in the GVRD*, Burnaby: Greater Vancouver Regional District.
Arthur, Brian (1989), 'Competing Technologies, Increasing Returns, and Lock-in by Historical Events', *Economic Journal*, **99**, 116–31.
Atkinson, D., A. Cristofaro, and J. Kolb (1991), 'Role of the Automobile in Urban Air Pollution', in Jefferson W. Tester, David O. Wood, and Nancy A. Ferrari (eds.), *Energy and the Environment in the 21st Century*, Cambridge, MA: MIT Press, pp.179–92.
Atkinson, Michael M. and William D. Coleman (1989), 'Strong States and Weak States: Sectoral Policy Networks in Advanced Capitalist Countries', *British Journal of Political Science*, **19** (1), 47–67.
Axelrod, R. (1984), *The Evolution of Cooperation*, New York: Basic Books.
Baar, Ellen (1992), 'Partnerships in the Development and Implementation of Canadian Air Quality Regulation', *Law and Policy*, **14** (1), 1–43.
Baar, Ellen (1995), 'Economic Instruments and Control of Secondary Air Pollutants in the Lower Fraser Valley', in Anthony Scott, John Robinson, and David Cohen (eds), *Managing Natural Resources in British Columbia: Markets Regulation and Sustainable Development*, Vancouver: UBC Press, pp.95–131.
Baar, Ellen (1996), 'Marrying Science and Policy', Chapter 5 of a manuscript in preparation, mimeo.
Bauer, M. and E. Cohen (1981), *Qui Gouverne les Groupes Industriels Français?*, Paris: Seuil.
Baumol, William J. and Wallace E. Oates (1988), *The Theory of Environmental Policy*, Cambridge: Cambridge University Press.
BC Hydro (1993), *The Burrard Utilization Study Report*, Vancouver: BC Hydro.
Bennett, Colin J. and Michael Howlett (1992), 'The lessons of Learning: Reconciling Theories of Policy Learning and Policy Change', *Policy Sciences*, **25** (3), 275–94.

Benoît, Bruno et al. (1994), *24 Maires de Lyon pour 2 Siècles d'Histoire*, Lyon: Lugd.
Bohn, Glenn (1996), 'Road Tolls a Tougher Sell in Suburbs', *Vancouver Sun*, 17 September.
Bonnel, P. (1994), 'Urban Car Policy in Europe', in *Car Free Cities?*, Amsterdam: Car Free Cities Conference, pp. 131–42.
Bonnet, J. (1975), *Lyon et son agglomération*, Paris: La Documentation Française, Notes et études documentaires No. 4207.
Bonnet, Jacques (1987), *Lyon et son Angglomération*, Paris: La Documentation Française.
Bovar-Concord Environmental (1995), *Economic Analysis of Air Quality Improvement in the Lower Fraser Valley*, Burnaby: Bovar-Concord.
Bressers, H., and J. Ligteringen (1997), *What To Do with Non-accessible Target Groups: Policy Strategies for Sustainable Consumption*, Presented at the 4th Meeting of the Concerted Action Network, 'The Ecological State,' Barcelona, 7–8 November 1997.
Brook, Jeffrey, Richard Burnett, Tom Dann, Daniel Krewski, and Renaud Vincent, (1995), 'Associations Between Ambient Particulate Sulfate and Admissions to Ontario Hospitals for Cardiac and Respiratory Diseases', *American Journal of Epidemiology*, **142** (1), 15–22.
Brook, J. R., Richard Burnett, S. Calmak, D. Krewski, H. Ozaynak, O. Philips, M. Raizenne, D. Steib, and R. Vincent (1988), 'The Association Between Ambient Carbon Monoxide Levels and Daily Mortality in Toronto, Canada', *Journal of Air and Waste Management Association*.
Buglione, E. (1996), 'Ecoautomobilismo ed Ecotasse: Prospettive e Concrete Realizzazionií', in C. Desideri (ed), *Qualitt dell'aria e Automobili*, Milan: Giuffrè, pp. 125–80.
Burnett, Richard, Robert Dales, and Mack E. Raizenne (1994), 'Effects of Low Ambient Levels of Ozone and Sulphates on the Frequency of Respiratory Admissions to Ontario Hospitals', *Environmental Research*, **65**, 271–90.
Campbell, Monica E., Beth A. Benson, and Marie A. Muir (1995), 'Urban Air Quality and Human Health: A Toronto Perspective', *Canadian Journal of Public Health*, **86** (5), 351–7.
Canterbery, Ray E. and A. Marvasti (1992), 'The Coase Theorem as a Negative Externality', *Journal of Economic Issues*, **26** (4): 1179–89.
Canterbery, Ray E. and A. Marvasti (1994), 'Two Coases or Two Theorems', *Journal of Economic Issues*, **28** (1): 218–26.
CEMT (1989), *Systèmes de couverture des coûts d'infrastructure routières urbains*, Table Ronde 97, Paris: OECD.
CEMT (1994), Les péages routiers urbains, Table Ronde 97, Paris: OECD.
Centre for Sustainable Transportation (1997), *Definition and Vision of Sustainable Transportation*, Toronto: Centre for Sustainable Transportation.

Ceruti, M. and C. Testa (1991), 'Gli Otto Peccati Mortali della Cultura Verde', *Micromega*, **3** (91), 7–25.

City of Toronto, Department of Public Health (1993), *Outdoor Air Quality in Toronto: Issues and Concerns*, Toronto: City of Toronto.

Coase, Ronald (1990), 'The Problem of Social Cost', *Journal of Law and Economics*, **3** (1), 1–44.

Coleman, William and Anthony Perl (1997), 'Internationalized Policy Environments and Policy Network Analysis: Directions for Future Research,' presented at panel 18-6 of the 93rd annual meeting of the American Political Science Association, Washington, DC, 31 August.

Coleman, W.D. and G. Skogstad, (1990a), 'Policy Communities and Policy Networks: a Structural Approach', in W.D. Coleman and G. Skogstad (eds), *Policy Communities and Public Policy in Canada*, Mississauga: Copp Clark Pitman.

Coleman, William D. and Grace Skogstad (1990b), *Policy Communities and Public Policy in Canada: A Structural Approach*, Toronto: Copp Clark Pitman.

Communication au Congrès de l'ATEC, *Intermodalité et Complémentarité des Modes de Transports*, LET: Lyon.

Corbin, A. (1982), *Le Miasme et la Jonquille*, Paris: Champs Flammarion.

COREP, Polytechnic of Turin, CELP and LSE (1994), *The Cultural and Economic Conditions of Decision-Making for the Sustainable City*, Report to DG XII of the EU.

Crandall, Robert W., Howard K. Gruenspecht, Theodore E. Keeler, and Lester B. Lave (1986), *Regulating the Automobile*, Washington, DC: The Brookings Institution.

CSST (Centro Studi Sistemi Trasporti) (1994), '*Studio sull'attuazione del Decreto 12.11.1992 del Ministero dell'Ambiente*', mimeo.

Daganzo, Carlos F. (1995), 'A Pareto Optimum Congestion Reduction Scheme', *Transportation Research*, **29** (2), 139–54.

David, Paul (1985), 'Clio and the Economics of QWERTY', *American Economic Review*, **75** (2), 332–7

David, Paul and Dominique Foray (1995), 'Dépendence du Sentier et Économie de L'Innovation: un Rapide Tour d'Horizon', *Revue d'Economie Industrielle*, hors-série *Economie Industrielle: Développements Récents*, 27–52.

DDE (Groupe de Travail DDE/Conseil Général/COURLY) (1993), *Schéma de Grandes Voiries de l'Agglomération Lyonnaise*.

De Serpa, Allan (1993), 'Pigou and Coase in Retrospect', *Cambridge Journal of Economics*, **17**, 27–50.

De Spot, Michel (1994), *Air Quality Management Bylaw Fees and Authorised Emissions in 1993*, Burnaby: Greater Vancouver Regional District Air Quality and Source Control.

Deakin, Elizabeth (1993), 'Policy Responses in the USA', in David Bannister and Kenneth Button (eds), *Transport, the Environment, and Sustainable Development*, London: E & FN Spon, 79–101.

Dente, B. (1989), 'Il Governo localeí', in G. Freddi (ed.), *Scienza dell Amministrazione e Politiche Pubbliche*, Rome: NIS, pp.123–69.

Dente, B. (ed.) (1995), *Environmental Policy in Search of New Instruments*, Dordrecht: Kluwer.

Dente, B., P. Knoepfel, R. Lewanski, S. Manozzi, and S. Tozzi (1984), *Il Controllo dell'inquinamento Atmosferico in Italia: Analisi di una Politica Regolativa*, Rome: Officina Edizioni.

Desideri, C. (1993), 'L'inquinamento Atmosferico e Acustico Nelle Citt: Cronaca di una Emergenza Istituzionalel', in C. Desideri (ed.), *L'inquinamento Atmosferico e Acustico Nelle Citt: dall'Emergenza all'intervento Ordinario*, Rome: ISR.

Desideri, C. (1996), 'Inquinamento Atmosferico e Traffico Veicolare: Caratteri Contenuti e Prospettive della Regolazione Comunitaria e Nazionalei', in C. Desideri (ed.), *Qualitt dell'aria e Automobili*, Milan: Giuffrè, pp.1–69.

Desideri, Carlo and Rudy Lewanski (1996), 'Improving Air Quality in Italian Cities: the Outcome of an Emergency Policy Style', mimeo.

Donati, A. (ed.) (1996), 'Stato di Attuazione dei Piani Urbani del Traffico', WWF, mimeo.

Dowding, K. (1995), 'Model or Metaphor? A Critical Review of the Policy Networks Approach', *Political Studies*, **43**, 136–58.

Downs, Anthony (1992), *Stuck in Traffic: Coping with Peak Hour Traffic Congestion*, Washington, DC: Brookings Institution.

Dunn, James A. and Anthony Perl (1996) 'Building the Political Infrastructure for High Speed Rail in North America', *Transportation Quarterly*, **50** (1), 5–22.

Eck, Theodore (1991), 'Positioning for the 1990s – The Amoco Outlook', in Jefferson W. Tester, David O. Wood, and Nancy A. Ferrari (eds), *Energy and the Environment in the 21st Century*, Cambridge: MIT Press, pp. 225–6.

Edelman, M. (1964), *The Symbolic Uses of Politics*, Urbana: University of Illinois Press.

Edelman, M. (1991), *Piecès et Règles du Jeu Politique*, Paris: Seuil.

Else, P.K (1981), 'The Theory of Optimum Congestion Taxes', *Journal of Transport Economics and Policy*, **15** (3), 217–32.

Elsom, D. (1996), *Smog Alert: Managing Urban Air Quality*, London: Earthspan.

Evans, Andrew (1992), 'Road Congestion Pricing: When it is a Good Policy?', *Journal of Transport Economics and Policy*, **26** (3), 213–43

Farmar-Bowers, Quentin (1996), 'Air Quality Programs in Vancouver, British Columbia, Canada', *Road and Transport Research*, **5** (2), 51–6.

Faure, A., G. Pollet and P. Warin (eds) (1995), *La Construction du Sens dans les Politiques Publiques*, Paris: L'Hanna.
Fiorillo, A. (1996), 'Ecosistema Urbanoí: Caos', *Quaderni di LegAmbiente*, 7 December.
Fischer, F. and J. Forester (eds) (1993), *The Argumentative Turn in Policy Analysis and Planning*, London: UCL Press.
Flink, James J (1988), *The Automobile Age*, London: MIT Press.
Foray, Dominique (1996), 'Diversité, Sélection et Standardisation: les Nouveaux Modes de Gestion du Changement Technique', *Revue d'Economie Industrielle*, **75**, 257–76.
French Administrative Directory (1970–74), Paris: Société du Bottin Administraif.
Fürsorge- und Gesundheitsdirektion der Stadt Bern (1992–), *Verwaltungsbericht*, Bern: Stadtkanzlei.
Garrett, G. and B. Weingast (1993), 'Ideas, Interests and Institutions: Constructing the European Community's Internal Market', in J. Goldstein and Robert Keohane (eds), *Ideas and Foreign Policy: Beliefs. Institutions and Political Change*, Ithaca: Cornell University Press, pp. 173–206.
Gemeinderat der Stadt Bern (1982), *Umwelt, Stadt und Verkehr: Kurzbericht zu den Verkehrskonzepten der Stadt Bern*, Bern: Stadtplanungsamt.
Gemeinderat der Stadt Bern (1983), *Umwelt. Stadt und Verkehr: Kurzbericht zur Parkraumplanung der Stadt Bern*, Bern: Stadtplanungsamt.
Gemeinderat der Stadt Bern (1992), *Entwurf zum Rämlichen Stadtentwicklungskonzept Bern*, Bern: Stadtplanungsamt.
Giuliani, M. (1989), 'Regolazione Senza Regole: Il Caso Italiano e le Possibili Interpretazionif', *Rivista Trimestrale di Scienza dell'Amministrazione*, **4**, 3–44.
Giuliani, M. (1997), *Decidere per l'Ambiente. Attori Processi, Politiche*, Milano: Istituto per l'Ambiente.
Giuliano, Genevieve (1992), 'An Assessment of the Political Acceptability of Congestion Pricing', *Transportation*, **19** (4), 335–53.
Goddard, Stephen D. (1994), *Getting There: The Epic Struggle Between Road and Rail in the American Century*, New York: HarperCollins.
Goldberg, Michael A. (1986), *The Myth of the North American City: Continentalism Challenged*, Vancouver: UBC Press.
Goldstein, J. and Robert Keohane (eds) (1993), *Ideas and Foreign Policy. Beliefs. Institutions and Political Change*, Ithaca: Cornell University Press.
Goodin, R.E. (ed.) (1996), *The Theory of Institutional Design*, Cambridge: Cambridge University Press.
Grant, Wyn (1996), 'Direct Democracy in California: Example or Warning?', *Democratization*, **3**, 133–49.

Greater Vancouver Regional District (GVRD) (1994), *Overview: GVRD Air-Quality Management Plan*, Burnaby: Greater Vancouver Regional District.

Greater Vancouver Regional District (GVRD) (1995a), *Liveable Region Strategic Plan*, Burnaby: Greater Vancouver Regional District.

Greater Vancouver Regional District (GVRD) (1995b), *Our Future*, Burnaby: Greater Vancouver Regional District.

Hall, P. (1986), *Governing the Economy: The Politics of State Intervention in Britain and France*, New York: Oxford University Press.

Hall, P. and R. Taylor (1996), 'Political Science and the Three New Institutionalisms', *Political Studies*, **44**, 936–57.

Harrison, Kathryn (1996), *Passing the Buck: Federalism and Canadian Environmental Policy*, Vancouver: UBC Press.

Hassenteufel, Patrick (1995), 'Do Policy Networks Matter? Lifting Descriptif et Analyse de l'Etat en Interaction', in Patrick LeGales and Mark Thatcher (eds), *Les Réseaux de politique publique. Debat autour des policy networks*, Paris: L'Harmattan.

Hayward, J. (1986), *The State and the Market Economy: Industrial Patriotism and Economic Intervention in France*, Brighton: Harvester Press.

Hösle, V. (1991), *Philosopie der Ökologischen Krise*, Munich: Oscar Beck.

House of Lords (1996), *First Report of the Select Committee on Science and Technology 1996–7, Towards Zero Emissions for Road Transport*, London: HMSO.

IFEN (1995), *L'environnement en France 1994–1995*, Paris: Dunod.

ISTAT (1993), *Statistiche Ambientali*, Rome.

ISTAT (1996), *Statistiche Ambientali*, Rome.

Istituto per l'Ambiente (1994), *Secondo Natura*, Milan: Rizzoli.

Jänicke, Martin (1978), 'Umweltpolitik im kapitalistischen Industriesystem. Eine einführende Problemskizze', in *Umweltpolitik: Beiträge zur Politologie des Umweltschutzes*, Opladen: UTB, pp. 9–35.

Jones, C.O. (1977), *An Introduction to the Study of Public Policy*, North Scituate: Duxbury Press.

Jordan, G. (1990), 'Policy Community Realism versus New Institutionalist Ambiguity', *Political Studies*, **38** (3), 470–84.

Jordan, Grant and Klaus Schubert (1992), 'A Preliminary Ordering of Policy Network Labels', *European Journal of Political Research*, **21** (1), 95–123.

Kato, J. (1996), 'Institutions and Rationality in Politics – Three Varieties of NeoInstitutionalists, *British Journal of Political Science*, **26**, 553–82.

Kenis, Patrick and Volker Schneider (1991), 'Policy Networks and Policy Analysis: Scrutinizing a New Analytical Toolbox', in Bernd Martin and Renate Mayntz (eds), *Policy Networks: Empirical Evidence and Theoretical Consideration*, Frankfurt am Main and Boulder: Campus, pp. 25–59.

KIGA (1991), *Massnahmenplan zur Luftreinhaltung in der Region Bern – Teilmassnahmenplan Verkehr*, Bern: Kantonales Amt für Industrie, Gewerbe und Arbeit, (KIGA).

King, G., R. Keohane, and S. Verba (1994), *Designing Social Inquiry*, Princeton: Princeton University Press.

Kingdon, J.W. (1984), *Agendas, Alternatives and Public Policies*, Boston: Little Brown.

Kissling-Näf, I. and Knoepfel, P. (1998), 'Social Learning in Policy Networks', *Policy and Politics*, **26** (3), 343–68.

Kitsuse, John I. and Malcolm Spector (1977), *Constructing Social Problems*, London: Cummings.

Knight, Franck (1924), 'Some Fallacies in the Interpretation of Social Costs' *Quarterly Journal of Economics*, **38**, 582–605.

Knoepfel, P. (1994), 'Learning Patterns in Swiss Environmental Policies', paper presented at the annual meeting of the American Political Science Association, New York.

Knoepfel, P. and C. Larrue (1985), 'Distribution Spatiale et Mise en Oeuvre d'une Politique Publique: le Cas de la Pollution Atmosphérique', *Politiques et Management Public*, (2), 43–70.

Knoepfel, Peter, Rita Imhof, and Willi Zimmermann (1995), *Luftreinhaltepolitik im Labor der Städte. Der Massnahmenplan – Wirkungen eines Neuen Instruments der Bundespolitik im Verkehr*, Basel and Frankfurt am Main: Helbing & Lichtenhahn.

Knoepfel, P., I. Kissling-Näf, and D. Marek (1997), *Lernen in offentlichen Politiken*, Basel and Frankfurt am Main: Helbing & Lichtenhahn.

Kraines, D. and V. Kraines (1993), 'Learning to Cooperate with Pavlov. An Adaptive Strategy for the Iterated Prisoner's Dilemma with Noise', *Theory and Decision*, **35** (2), 107–49.

Lanzalaco, L., R. Lizzi, and L. Martinelli (1996), 'Politiche Pubbliche per il Governo delle area Metropolitane nei Settori dell'inquinamento Acustico e Atmosferico', *CNR Report* (ed. G. Freddi), Rome: CNR.

Lascoumes, P. (1994), *L'Éco-pouvoir; Environnement et Politiques*, Paris: De La Découverte.

LegAmbiente (1995), *Ambiente Italia 1995*, Milan: Edizioni Ambiente.

LegAmbiente (1996), *Ambiente Italia Rapporto Sullo Stato del Paese a Confronto con l'Europa*, Milan: Edizioni Ambiente.

LegAmbiente (1997), *Ambiente Italia 1997*, Milan: Edizioni Ambiente.

Lembruch, Gerhard and Philipe Schmitter (eds) (1982), *Patterns of Corporatist Policy Making*, Beverly Hills: Sage.

Lewanski, R. (1990), 'La Polotica Ambientale', in B. Dente (ed.), *Le Polotoche Pubbliche in Italia*, Bologna: Il Mulino, pp. 281–314.

Lewanski, R. (1997), *Governare l'Ambiente: Attori e Processi Delta Politica Ambientale*, Bologna: II Mulino.
Liberatore, A. and R. Lewanski (1990), 'The Evolution of Italian Environmental Policy', *Environment*, **32** (5), 10–40.
Lipfert, Frederick W. and Ronald E. Wyzga (1995), 'Air Pollution and Mortality: Issues and Uncertainties' in *Journal of the Air & Waste Management Association*, **45** (12), December, 949–66.
Lizzi, R. (1997), 'La Politique Italienne du Contrôle de la Pollution Atmosphérique: Limites, Acteurs et Réseaux des Policy', *Pôle Sud-Revue de Science Politique*, (6), 86–100.
Lizzi, R., L. Martinelli, and L. Lanzalaco (1996), 'Rapporto di Ricerca al CNR,' mimeo.
Loi no 96-1236 du 30 Décembre 1996, 'Loi sur l'Air et l'Utilisation Rationnelle de l'Énergie', *Journal Officiel*, 1/1/1997.
Lojkine, Jean (1974), *La Politique Urbaine dans la Région Lyonnaise, 1945–1972*, Paris: Mouton.
Long, Ross A. (1993), *Transportation Demand Measures and Their Potential for Application in Greater Vancouver*, Burnaby: Greater Vancouver Regional District.
Lundquist, L. (1980), *The Hare and the Tortoise: Clean Air Policies in the United States and Sweden*, Ann Arbor: University of Michigan Press.
MacKenzie, J., R. Dower, and P.T. Chen (1992), *The Going Rate: What It Really Costs to Drive*, Washington, DC: World Resources Institute.
Magnusson, Warren (1983), 'Toronto', in Warren Magnusson and Andrew Sancton (eds), *City Politics in Canada*, Toronto: University of Toronto Press, pp. 94–139.
Majone, Giandomenico (1989), *Evidence, Argument and Persuasion in the Policy Process*, New Haven: Yale University Press.
March, J.G. and J.P. Olsen (1984), 'The New Institutionalism: Organizational Factors in Political Life', *American Political Science Review*, **78**, 734–49.
March, J.G. and J.P. Olsen (1989), *Rediscovering Institutions: The Organizational Basis of Politics*, New York: The Free Press.
March, J.G. and J.P. Olsen (1995), *Democratic Governance*, New York: The Free Press.
Marek, Daniel (1995), *Luftreinhaltung und Verkehr in der Stadt Bern. Akteurnetzwerke der Verkehrsberuhieung und Förderuy Uwltfreundlicher Verkehrsmittel an vier Beispielen*, Lausanne: Cahiers de l'IDHEAP, No. 143.
Marin, Bernd and Renate Mayntz (1991), 'Studying Policy Networks', in Bernd Marin and Renate Mayntz (eds), *Policy Networks: Empirical Evidence and Theoretical Considerations*, Frankfurt am Main and Boulder: Campus, pp.11–23.

Marlot, Grégoire (1995), *Politique Urbaine et Congestion: le Politicien, l'Ingénieur et l'Usager*, Mémoire de DEA sous la Direction de C. Raux, Université Lumière Lyon 2.

Marlot, Grégoire (1996), 'Institutions et régulation de la congestion', Working Paper Laboratoire d'Economie des Transports, Lyon, No. 34.

Marzano, M. (forthcoming), 'Le Politiche di Limitazione del Traffico', in L. Bobbio and F. Ferraresi (eds), *Decidere in Comune, Analisi e Riflessioni su Sento Decisioni Comunali*, Bologna: Il Mulino.

May, A. (1992), 'Road Pricing: an International Perspective', *Transportation*, **19** (4), 313–33.

May, A., M. Roberts, and P. Mason (1992), 'The Development of Transport Strategies for Edinburgh', *Transportation*, **95**, 51–60.

Mayntz, Renate (ed.), *Policy Networks: Empirical Evidence and Theoretical 'deration'*, Frankfurt am Main and Boulder: Campus, pp.1–23.

Mazza, L. and Y. Rydin (eds) (1997), 'Urban Sustainability: Discourses, Networks and Policy Tools', *Progress in Planning*, 47.

Medema, Steve (1994), 'The Myth of Two Coases: What Coase is Really Saying', *Journal of Economic Issues*, **28** (1), 208–17.

Mennell, Morris (1995), 'Air Quality Planning in One of North America's Fastest Growing Regions', *Environmental Manager*, **1**, 22–6.

Merton, Robert K. and Patricia L. Kendall, (1979), 'Des Fokussierte Interview', in Christel Hopf and Elmar Weingarten (eds), *Qualitative Sozialforschung*, Stuttgart: Klett-Cotta, pp. 171–204.

Miller, P. and J. Moffett (1993), *The Price of Mobility: Uncovering the Hidden Costs of Transportation*, New York: National Resources Defence Council.

Ministero dei Trasporti e della Navigazione (1995a), *Conto Nazionale dei Trasporti*, Rome.

Ministero dei Trasporti e della Navigazione (1995b), *I Trasporti in Italia*, Rome.

Ministero dell'Ambiente (1989), *Relazione Sullo Stato dell'Ambiente*, Rome.

Ministero dell'Ambiente (1992), *Relazione Sullo Stato dell'Ambiente*, Rome: Istituto Poligrafico e Zecca dello Stato.

Ministero dell'Ambiente (1996), *Lo Stato dell'Ambiente in Italia. Sintesi della Terza Relazione al Parlamento*, Rome: Istituto Poligrafico e Zecca cello Stato.

Municipality of Metropolitan Toronto (1996), *Blue Ribbon Committee Report. Bad Air Alert: It's Killing Us*, Toronto: Municipality of Metropolitan Toronto.

Newman, P.W.G. and J. R. Kenworthy (1988), 'The Transport–Energy Trade-off: Fuel Efficient Traffic Versus Fuel Efficient Cities', *Transportation Research*, 22 A, 205–21.

Newman, P.W.G. and J.R. Kenworthy (1989), *Cities and Automobile Dependence: A Sourcebook*, Aldershot: Gower.

Niskanen, W.A. (1971), *Bureaucracy and Representative Government*, Chicago: Rand McNally.
North, D. (1990), *Institutions, Institutional Chance and Economic Performance*, Cambridge: Cambridge University Press.
Nullmeier F. (1993), 'Wissen und Policy: Forschung. Wissenspolitologie und Rhetorisch-Dialektisches Handlungsmodell', *Politische Vierteljahresschrift Sonderheft*, **34**, 175–96.
OECD (Organisation of Economic Co-operation and Development) (1994), *Environmental Performance Review: Italy*, Paris: OECD.
Official Journal (1903), 14/6/1903.
Official Journal (1991), Senate, 6/7/1961, p.717.
Offner, J.M. (1990), 'Configurations du pouvoir technico-politique local et diversification des compétences dans le secteur des transports collectifs urbain', rapport INRETS Nol. 98, Arcueil: INRETS.
Olson, M. (1965), *The Logic of Collective Action: Public Goods and the Theory of Groups*, Cambridge, MA: Harvard University Press.
Ontario Ministry of Environment and Energy (OMOEE). (1992), 'MOE Highlights', in *Clean Talk*, March/April, Toronto: OMOEE.
Ontario Ministry of Environment and Energy (OMOEE) (1995), *Air Quality Report in Ontario: 1994 Comprehensive Report*, Toronto: Queen's Printers for Ontario.
Ostrom, E., R. Gardner, and J. Walker (1994), *Rules, Games, and Common Pool Resources*, Ann Arbor: University of Michigan Press.
Ozkaynak, A. et al. (1995), 'Association between Daily Mortality and Air Pollution in Toronto, Canada', *Proceedings of the International Society for Environmental Epidemiology*, Noordwijkerhout, The Netherlands.
Pappi, Franz U. (1987), 'Die Netzwerk-Analyse aus Soziologischer Sicht', in *Methoden der Netzwerkanalyse: Techniken der Empirischen Sozialforschung* 1, München: Oldenbourg.
Perl, Anthony and J.D Han (1994), 'Automotive Pricing and Sustainable Mobility', in *World Transport Research*, Proceedings of the First World Conference on Transportation Research, 171–5.
Perl, Anthony and J.D. Han (1995), 'Evaluating the Environmental Component of Automobile Pricing Schemes', in *World Transport Research*, Proceedings of the 7th World Conference on Transport Research, 265–75.
Perl, Anthony and J.D. Han (1996), 'Evaluating the Environmental Component of Automobile Pricing Schemes', in D. Hensher, J. King, and T. H. Oum (eds), *World Transport Research*, Oxford: Elsevier Science Ltd.
Perl, Anthony and J. Hargraft (1996), 'When Policy Networks Collide: The Institutional Constraint on Air Pollution Strategies in Two Canadian Cities, mimeo.

Perl, Anthony and John Pucher (1995), 'Transit in Trouble? The Policy Challenge Posed by Canada's Changing Urban Mobility', *Canadian Public Policy*, **22** (3), 261–83.
Peters, G.B. (1996), 'Political Institutions, Old and New', in R.E. Goodin and H.D. Klingemann (eds), *A New Handbook of Political Science*, Oxford: Oxford University Press, pp. 205–20.
Pierson, P. and K. Weaver (1993), 'Imposing Losses in Pension Policy', in K. Weaver and B. Rockman (eds), *Do Institutions Matter? Government Capabilities in the United States and Abroad*, Washington, DC: The Brookings Institution, pp. 110–50.
Polizeidirektion der Stadt Bern (1991), '*Probleme, Prioritäten im Bereich Verkehr*', Bern: unpublished paper.
Ponti, M. and Vittadini M.R. (1990), 'Italy', in J.P. Barde and K. Button (eds), *Transport Policy and the Environment*, London: Earthscan.
Präsidialdirektion der Stadt Bern (1987–1992), *Statistisches Jahrbuch der Stadt Bern*, Bern: Amt für Statistik.
Pucher, J. and C. Lefêvre (1996), *The Urban Transport Crisis in Europe and North America*, London: Macmillan.
Putnam, R. (1993), *Making Democracy Work*, Princeton: Princeton University Press.
Radaelli, C.M. (1995), 'The Role of Knowledge in the Policy Process', *Journal of European Public Policy*, **2** (2), 159–83.
Raux, Charles (1994), 'Le Peage Urbain: une Incitation au Changement de Mode de Transport', in *Communication au Congrès de l'ATEC Intermodalité et Complémentarité des Modes de Transports*, Lyon: LET.
Raux, Charles and Martin Lee-Gosselin (eds) (1991), *La Mobilité Urbaine: de la Paralysie au Péage?*, Lyon: Centre Jacques Cartier.
Raux, Charles, O. Anden, B. Faive O'Arcier, and C. Godinot (1995), *Les Réactions au Péage Urbain. Enquête Exploratoire*, Lyon: LET, Etudes et Recherches.
Rhodes, R.A.W. and D. Marsh (1992), 'Policy Networks in British Politics: a Critique of Existing Approaches', in D. Marsh and R.A.W. Rhodes (eds), *Policy Networks in British Government*, Oxford: Clarendon Press, pp. 1–26.
Richardson, H.W. (1978), *Regional and Urban Economics*, Harmondsworth: Penguin Books.
Rietveld, P. (1996), 'Tradable Permits: Their Potential in the Regulation of Road Transport Externalities', working paper, Tinbergen Institute.
Rochefort, D.A. and R.W. Cobb (eds) (1994), *The Politics of Problem Definition. Shaping the Policy Agenda*, Lawrence: University Press of Kansas.
Roqueplo, P. (1988), *Pluies Acides: Menaces pour l'Europe*, Paris: CPE, Economical

Rose, Richard (1993), *Lesson-Drawing in Public Policy: A Guide to Learning Across Time and Space*, New Jersey: Chatham House.
Rothengatter, Werner (1992), 'Externalities of Transport,' mimeo.
Royal Commission on Environmental Pollution (1994), *Eighteenth Report. Transport and the Environment*, London: HMSO.
Rusk, James (1997), 'Ontario Makes Emission Testing For Vehicles Mandatory: Drive Clean Program, Which Will Get Under Way Next Summer, Aims To Cut Down On Pollutants That Cause Dangerous Smog', *Globe and Mail*, 23 August.
Sabatier, P. and H. Jenkins-Smith (1993), *Policy Change and Learning: An Advocacy Coalition Approach*, Boulder: Westview Press.
Sartori, G. (1987), *The Theory of Democracy Revisited*, Chatham: Chatham House.
Scharpf, F.W. (1993), 'Co-ordination in Hierarchies and Networks,' in F.W. Scharpf (ed.), *Games in Hierarchies and Networks*, Frankfurt am Main: U.A.P., pp. 125–65.
Schenkel, Walter (1995), 'Clean Air and Transport Policy in the Canton of Basle-Town,' COST-CITAIR final report.
Schneider, Volker (1988), *Politiknetzwerke der Chemikalienkontrolle: Analyse einer Transnationalen Politikentwicklune*, Berlin and New York: Walter de Grnyter.
Signorino, M. (ed.) (1996), *Ventanni di Politica Ambientale in Italia*, Rimini: Maggioli.
Skocpol, Theda (1985), 'Bringing the State Back In: Strategies of Analysis in Current Research', in Peter B. Evans, Dietrich Rueschemeyer, and Theda Skocpol (eds), *Bringing the State Back In*, Cambridge: Cambridge University Press, pp. 3–37.
Skou, Andersen M. (1994), *Governance by Green Taxes: Making Pollution Prevention Pay*, Manchester: Manchester University Press.
Small, Kenneth (1992), 'Using the Revenues of Congestion Pricing', *Transportation*, **19** (4), 359–81.
Spector, M. and J.I. Kituse (1987), *Constructing Social Problems*, New York: Aldine de Gruyer.
Sperling, Daniel (1991), 'An Incentive-Based Transition to Alternative Transportation Fuels', in Jefferson W. Tester, David O. Wood, and Nancy A. Ferrari (eds), *Energy and the Environment in the 21st Century*, Cambridge, MA: MIT Press, pp. 251–64.
Steinmo, S., K. Thelen, and F. Longstreth (eds) (1992), *Structuring Politics: Historical Institutionalism in Comparative Analysis*, Cambridge: Cambridge University Press.

Stone, D. (1988), *Policy Paradox and Political Reason*, Glenview: Scott Foresman.
Swiss Agency for the Environment, Forests and Landscape (1997), *Climate in Danger: Facts and Implications of the Greenhouse Effect*, Bern: Swiss Agency for the Environment, Forests and Landscape.
SYTRAL (1986), 'Plan de Déplacements Urbains: le Diagnostic', Dossier Technique Préparatoire.
Tennant, Paul and David Zirnhelt (1973), 'Metropolitan Government in Vancouver: The Strategy of Gentle Imposition', *Canadian Public Administration*, **16** (1), 124–38.
Tuohy, Carolyn Hughes (1992), *Policy and Politics in Canada: Institutionalized Ambivalence*, Philadelphia: Temple University Press.
UITP-IVECO (1994), Figure in *La Nuova Ecologia*, 2, p. 11.
Unione Petrolifera (1996), 'Informazioni Petrolifere', supplement to *Notizie Statistiche Petrolifere*, February.
Vancouver Board of Trade (1991), *Industrial Emission Reductions in the Lower Mainland: Report of the Vancouver Board of Trade Environmental Task Force*, Vancouver: Vancouver Board of Trade.
Vancouver Sun (1996), 'Crossed Wires: Planners Want More Buses as BC Transit Plans Service Cuts', *Vancouver Sun*, 17 September.
Verhoef E., P. Nijkamp, and P. Rietveld (1995), 'Second Best Regulation of Road Transport Externalities', *Journal of Transport Economics and Policy*, **29** (2), 147–67.
Verhoef, E., P. Nijkamp, and P. Rietveld (1996), 'Tradable Permits: Their Potential in the Regulation of Road Transport Externalities', Tinbergen Institute, working paper.
Verhoef, Erik (1994), 'External Effects and Social Costs of Road Transport', *Transportation Research*, **28** (4), 273–87.
Von Beyme, Klaus (1980), *Interessengruppen in der Demokratie*, München: Piper.
Von Weisäcker, E.U. (1994), *Earth Politics*, London: Zed Books.
Weaver, R. Kent (1986), 'The Politics of Blame Avoidance', *Journal of Public Policy*, **6** (4), 371–98.
Weaver, R. Kent. and B.A. Rockman (eds) (1993), *Do Institutions Matter?*, Washington, DC: The Brookings Institution.
Weiss, C. (1979), 'The Many Meanings of Research Utilization', *Public Administration Review*, **39** (5), 426–31.
World Commission on Environment and Development (1987), *Our Common Future*, Oxford: Oxford University Press.
Wright, Maurice (1988), 'Policy Community, Policy Network, and Comparative Industrial Policies', *Political Studies*, **36**, 593–612.

Yago, G. (1984), *The Decline of Transit: Urban Transportation in German and US Cities*, 1900-1970, Cambridge: Cambridge University Press.
Zeppetella, A. (1996), *Retorica per L'Ambiente*, Milan: F. Angeli.
Zimmermann, W., S. Wyss, and P. Neuenschwander (1997), *Luftreinhaltekampagnen Zurich und St. Gallen*, Chavannes: Cahier de l'IDHEAP.
Zolea, S. (1996), 'Problem Scientifici e Tecnici dell'Inquinamento Atmosferico: dalla Discussions Scientifica Agli Interventi Operativií, in C. Desideri (ed.), *Qualitt dell'Aria e Automobili*, Milan: Giuffrè, pp. 71–124.

Index

air pollution
    automobile and 15, 54, 98, 106, 107, 122, 123, 131, 148
    health effects of 1–2, 16
    measurement of 62–3, 103–4
    point sources of 20, 23, 156
    see also benzene; carbon monoxide; lead; nitrogen oxides; particles; sulphur oxides; volatile organic compounds
air quality management agencies 154–5
Allegrini, L. 71
Anderson, W.P. 14
Atkinson, D. 16, 18
Atkinson, M. 29
automobile industry 18–19, 94, 97, 98, 101
Axelrod, R. 81

Baar, E. 21, 23, 26, 27
Baumol, W.J. 123
belief systems 10, 11, 144
Bennett, C.J. 30
Benôit, B. 108, 111
benzene 2, 55, 56, 57, 64, 65, 72
Berne 12, 127–43, 147, 167
Bohn, G. 25
Bologna 39, 60, 61, 62, 63, 65–6, 74
Bonnel, P. 75
Bonnet, J. 111
Bovar-Concord Environmental 22, 27
Bressers, H. 163
budgetary problems 149
    fiscal centralization in Italy 69
Buglione, E. 61, 72
bureaucracy
    limitations of 42, 43, 45, 68, 72
    technical experts in 17, 87, 115–17
Burnett, R. 2, 16

business interests 9, 64, 65, 66, 76, 88, 89, 108, 110, 120, 136, 143, 156, 157

California 25, 26, 159
Canada 6, 13–30
    see also Toronto; Vancouver
car parking *see* parking
carbon monoxide 54, 55, 106, 107, 122
Ceruti, M. 75
Coase, R. 123
Coleman, W.D. 4, 11, 13, 22, 29, 117
congestion 10, 73, 78, 91, 101–2, 122, 147–8, 152
corporatism 160
courts 57, 65, 165
    role of in Italy 69–70, 166
Crandall, R.W. 18
cyclists 135, 155

De Spot, M. 23
Deakin, E. 16
Dente, B. 32, 67, 69, 72
Dowding, K. 4
Dunn, J.A. 29

Eck, T. 18
economic instruments 75, 161, 162, 163, 164
Edelman, M. 66, 93
Elsom, D. 2
enforcement *see* implementation; police, enforcement and
environmental groups *see* green movement
equity issues 145–6, 160
European Union 10, 55, 57, 70, 100, 101, 102–3 105–6, 149,163

Farmar-Bowers, Q. 23
Faure, A. 49

federalism 12, 14, 15, 69, 131
Fiorillo, A. 63
fiscal crisis see budgetary problems
Fischer, F. 33
Flink, J.J. 18
Florence 53, 59, 61, 62, 65, 67, 74
Forester, J. 33
France 6, 7, 10–12, 93–126, 152
  see also Lyon, Paris

Garrett, G. 31
Goddard, S.D. 18
Goldberg, M. 15
Goldstein, J. 50
Goodin, R.E. 31
green movement 10, 40, 62, 64, 73–4, 75, 87, 89, 130–31, 148

Hall, P. 4, 31
Han, J.D. 123, 148
Harrison, K. 15, 19
Hassenteufel, P. 4
Hösle, V. 80
Howlett, M. 30

Imhof, R. 128, 133, 142, 149
implementation 3
  conditions for successful 14, 27, 36, 49, 89–91, 142, 162
  problems of in Italy 67–8
institutions
  definition of 32
  new frameworks for 157–60
institutionalization
  definition of 46–7
  measurement of 47
  need for 160
issue networks 5
Italy 6, 8–10, 11, 31–92, 147, 163
  see also Bologna, Florence, Milan, Modena, Turin

Jänicke, M. 142
Jenkins-Smith, H. 71, 143, 166, 167
Jones, C.O. 93
Jordan, G. 4, 32, 50, 127
judiciary see courts

Kato, J. 31
Kendall, P.L. 129

Kenis, P. 4, 127
Kenworthy, J.R. 119, 120, 148, 156
Keohane, R. 37, 50
King, G. 37
Kingdon, J.W. 91, 92
Kissling-Näf, I. 166, 167
Kituse, J.I. 93
Kraines, D. 92
Kraines, V. 92

Lanzalaco, L. 51
lead 16, 54–5, 103, 107, 122
Lefêvre, C. 52, 61
Lehmbruch, G. 160
Liberatore, A. 36
Ligteringen, J. 163
Lipfert, F.W. 2
Lizzi, R. 36, 53, 66, 68
Long, R.A. 22
Lojkine, J. 108
Longstreth, F. 31
Lundquist, L. 32
Lyon 11, 12, 107–26, 146, 148, 150, 156, 157, 158

MacKenzie, J. 1
Magnusson, W. 15
Majone, G. 29, 50
March, J.G. 4, 31, 160
Marsh, D. 4, 5
Marzano, M. 73
Mayntz, R. 127
Mazza, L. 74
media 64–5, 67, 71, 107–8, 120
Mennell, M. 16, 23
Merton, R.K. 129
Milan 39, 53, 54, 59, 60, 62, 67, 72, 74, 87
Miller, P. 1
Modena 56, 60, 63, 66, 67
Moffett, J. 1

new institutionalism 4, 31
Newman, P.W.G. 119, 120, 148, 156
Niskanen, W.A. 115
nitrogen oxides 16, 54, 55, 100–101, 103, 107, 119, 122, 131
North, D. 31
Nullmeier, F. 167

Oates, W.E. 123
Offner, J.M. 115
ozone 16, 55, 100, 131
    EU directive on 106
Olsen, J.P. 4, 31, 160
Olson, M. 66
Ostrom, E. 153

Pappi, F.U. 129
Paris 100, 102, 107
parking
    introduction of charges for in Italy 9, 60–61, 76–8, 89, 91
    owners of parking space 154
    residents' permits in Switzerland 140–41
participation 43, 49, 165
see also referendums
particles 2, 16, 54, 55, 107, 122
pedestrians 155
Peters, G.B. 31
Pierson, P. 56
police
    competence of 65
    core actors in policy networks 12, 140, 141, 157
    enforcement and 68, 90, 133–4, 167
    health concerns of 65, 68
    violence against 65
policy community 13, 71, 147, 151–2, 155, 160, 163–4, 165, 166, 167
policy learning 9, 14, 27–30, 71, 164–7
policy networks 4–7, 10, 13
    composition of 10, 11, 96, 102, 103, 105, 111, 114–15, 117, 128, 138–9, 140, 144–5, 152, 167
    definitions of 5, 13, 167
    encompassing 20, 28–30, 48, 149, 155, 160, 164–5
    functions of 49
    link with cognitive models 49
    number of 15–19, 28
    policy effectiveness and 8, 12, 13–14, 20, 28–30, 127–8, 142, 158–9
    problem definition and 10, 91, 93, 95, 98, 105, 152, 155
    size of 12, 48, 103, 128, 138, 142, 144–5, 152
policy transfer 26

policy web 64, 65
political parties
    influence of in Italy 88
    policy impact of 17, 19, 24, 25, 130–31
    see also green movement
Pollet, G. 49
Ponti, M. 55, 69, 72
prisoner's dilemma model 81–3, 91–2
property rights 150, 153–4
public participation see participation
public transport 19, 121, 150–51
    neglect of in Lyons 108–9, 111, 112–14
    operators of 154, 155–6
    use of in Italy 61–2
Pucher, J. 19, 52, 61
Putnam, R. 34

Radaelli, C. 39
referendums 66–7, 74, 87, 159, 165
regulation 32, 123–4, 161–3
Rhodes, R. 4, 5
Richardson, H.W. 92
road pricing 14, 25, 26, 61, 122, 123, 161, 162
    see also economic instruments
Rockman, B.A. 31
Rome 53, 54, 55, 60, 61, 65, 72
Roqueplo, P. 106
Rose, R. 26
Rusk, J. 16
Rydin, Y. 74

Sabatier, P. 71, 166, 167
Sartori, G. 51
Scharpf, F.W. 5, 164
Schenkel, W. 126
Schmitter, P. 160
Schneider, B. 4, 127, 129
Schubert, K. 4, 50, 127
Signorio, M. 36
Skocpol, T. 21
Skogstad, G. 11, 13, 22, 117
Skou Andersen, M. 172
Small, K. 123, 155
smog precursors, see ozone, volatile organic compounds
social exclusion 150, 156

soft institutions 3, 8
  benefits of 32
  definition of 32
  limits of 160
Spector, M. 93
Sperling, D. 18
state
  capacity of 21, 29–30, 151
  role of 153–4
Steinmo, S. 31
Stone, D. 50
sulphur oxides 16, 55, 99, 100, 103, 122, 131
sustainable development 145–6, 153, 159–60
Switzerland 6–7, 12, 127–43, 159, 161, 163–4
  neighbourhood associations in 136
  see also Berne

Taylor, R. 31
Tennant, P. 22
Testa, C. 75
Thelen, K. 31
Toronto 6, 7–8, 13, 14, 15, 16–20, 28–9
Tuohy, C.H. 15
Turin 53, 57, 60, 62, 73–92, 147, 149, 150, 162, 163, 166

United States 15, 92, 131
  see also California

Vancouver 6, 7–8, 11, 13–15, 20–29, 143, 147, 156, 158, 160, 167
vehicle inspection and maintenance programmes 16, 25–6, 58
Verba, S. 37
Vittadini, M.R. 55, 69, 72
volatile organic compounds 54, 107, 122, 131
Von Beyme, K. 143

Warin, P. 49
Weaver, R.K. 30, 31, 56
Weingast, B. 31
Weiss, C. 44
Woudsma, C. 14
Wright, M. 13
Wyzga, R.E. 2

Yago, G. 108

Zeppetella, A. 75, 81
Zimmermann, W. 128, 133, 142, 149, 161, 164
Zirnhelt, D. 22
Zolea, S. 63, 64, 72